Reminiscences of Ahmed H. Zewail

Photons, Electrons and What Else?

A Portrait from Close Range.
Remembrances of his Group Members and Family

Reminiscences of Ahmed H. Zewail

Photons, Electrons and What Else?

A Portrait from Close Range.
Remembrances of his Group Members and Family

Editors

Abderrazzak Douhal
Universidad de Castilla la Mancha, Spain

John Spencer Baskin
Caltech, USA

Dongping Zhong
The Ohio State University, USA

World Scientific

NEW JERSEY • LONDON • SINGAPORE • BEIJING • SHANGHAI • HONG KONG • TAIPEI • CHENNAI • TOKYO

Published by

World Scientific Publishing Co. Pte. Ltd.
5 Toh Tuck Link, Singapore 596224
USA office: 27 Warren Street, Suite 401-402, Hackensack, NJ 07601
UK office: 57 Shelton Street, Covent Garden, London WC2H 9HE

Library of Congress Cataloging-in-Publication Data
Names: Zewail, Ahmed H. | Douhal, Abderrazzak, editor. | Baskin, John Spencer, editor. | Zhong, Dongping, editor.
Title: Reminiscences of Ahmed H. Zewail : photons, electrons, and what else? : a portrait from close range : remembrances of his group members and family / editors, Abderrazzak Douhal (Universidad de Castilla La Mancha, Spain), John Spencer Baskin (Caltech, USA), Dongping Zhong (The Ohio State University, USA).
Description: New Jersey : World Scientific, 2018.
Identifiers: LCCN 2017053356| ISBN 9789813231535 (hardcover : alk. paper) | ISBN 981323153X (hardcover : alk. paper) | ISBN 9789813231658 (pbk. : alk. paper) | ISBN 9813231653 (pbk. : alk. paper)
Subjects: LCSH: Zewail, Ahmed H. | Femtochemistry. | Laser photochemistry. | Photochemistry. | Chemical bonds. | Chemical kinetics.
Classification: LCC QD716.L37 Z49 2018 | DDC 540.92--dc23
LC record available at https://lccn.loc.gov/2017053356

British Library Cataloguing-in-Publication Data
A catalogue record for this book is available from the British Library.

For any available supplementary material, please visit
http://www.worldscientific.com/worldscibooks/10.1142/10750#t=suppl

Preface

The project of making this book on Reminiscences of Ahmed started a few weeks after his sudden death. While we were too shocked and unable to write in newspapers and scientific journals anything about Ahmed's legacy, we started with the idea of making the book, involving as many previous and current members from Ahmed's scientific family as possible. At the same time, a book focused on the scientific legacy of Ahmed was already in progress, being edited and written by Ahmed's colleagues. We agreed that our book should be different, written only by members of Ahmed's scientific family, and focused on personal stories to develop science and related technology at Femtoland. By the middle of November 2016, we drafted the plan and started e-mailing our colleagues who had worked with Ahmed. Many of the e-mail addresses used were not up-to-date, but we got help from Maggie, De Anne and other colleagues to correct the e-mail address list. Many of our colleagues welcomed the idea and expressed their positive interest in contributing to the book and some others did not reply (we maybe never reached them). By the end of June 2017, we got more than 50 contributions expressing very nice, touching, and funny personal stories from the early days of Ahmed's walks at Caltech, until the middle of 2016. The book also contains many pictures of Ahmed with the authors. Reading these stories from his Ph.D. students, postdoctoral researchers and senior visiting scholars gives another dimension to Ahmed as a scientist and a "boss" making and discussing research plans, results, and correcting and writing papers.

The contributions (Chapters) in the book are in a chronological order which we hope will allow getting the feeling of how Ahmed's concerns in science evolved from his first days at Caltech using picosecond technology to later femtosecond spectroscopy to elucidate gas-phase reaction dynamics to his fight to develop ultrafast electron microscopy to explore materials and biological systems, and of course passing by the Nobel Prize award period. By making this unique book of reminiscences, we also express our sincere gratitude and our thanks to Ahmed, the scientist, the man, and the friend. The first two chapters are from Maha and Nabeel Zewail, a daughter and a son of Ahmed, sharing with us intimate stories with their father and his thinking.

We hope, dear reader, that you enjoy this book, not only on science, but also on making science with a great scientist.

Abderrazzak Douhal
John Spencer Baskin
Dongping Zhong

Contents

List of Contributors

Fida F. Al-Adel
KFUPM BOX # 732
Dhahran 31261, Saudi Arabia

Luis Bañares
Departamento de Química Física I
Facultad de Ciencias Químicas
Universidad Complutense de Madrid
28040 Madrid, Spain

Spencer Baskin
312 Hawk Lane
Glendora, CA 91741, USA

Thomas Baumert
Universitaet Kassel
Institut fuer Physik
Kassel, Germany

Thorsten M. Bernhardt
Institut für Oberflächenchemie und Katalyse
Universität Ulm
D-89069 Ulm, Germany

Fabrizio Carbone
Laboratory for Ultrafast Microscopy
and Electron Scattering (LUMES)
Institute for Condensed Matter Physics
Ecole Polytechnique Féderale de Lausanne
CH-1015 Lausanne, Switzerland

Bin Chen
School of Chemistry and Chemical Engineering
Shanghai Jiao Tong University
Shanghai, China

Po-Yuan James Cheng
Department of Chemistry
National Tsing Hua University
Hsinchu 30043, Taiwan

Marcos Dantus
Department of Chemistry and Department of Physics and Astronomy
Michigan State University
East Lansing, MI 48824, USA

Daniel R. Dawson
818 Liberty St.
Ashland, Oregon 97520, USA

Steven De Feyter
Department of Chemistry
University of Leuven
Leuven, Belgium

Eric Wei-Guang Diau
Department of Applied Chemistry and Institute of Molecular Science
National Chiao Tung University
Hsinchu 30010, Taiwan

Abderrazzak Douhal
Facultad de Ciencias Ambientales y Bioquimica
Universidad de Castilla La Mancha
Avenida Carlos III, S.N.
45071 Toledo, Spain

Jon Feenstra
WINGS Birding Tours
Director, Woodstar Biological, LLC
Altadena, California, USA

Torsten Fiebig
Beckman Laser Institute and Medical Clinic
University of California, Irvine
1002 Health Sciences
Irvine, CA 92612, USA

Xuewen Fu
Brookhaven Matter Physics & Materials Science Department
Upton, New York, USA

Roberto A. Garza-López
Department of Chemistry
Pomona College
Claremont, California, USA

Boyd Goodson
Department of Chemistry and Biochemistry
Southern Illinois University
Carbondale, Illinois, USA

Martin Gruebele
Department of Chemistry
University of Illinois at Urbana-Champaign
505 South Mathews Avenue
Urbana, IL 61801, USA

Hua Guo
Department of Chemistry and Chemical Biology
Department of Physics and Astronomy
University of New Mexico
Albuquerque, New Mexico, USA

Michael Gutmann
LIOP-TEC GmbH, Industriestrasse 4
Radevormwald, Germany

Niels Engholm Henriksen
Department of Chemistry
Technical University of Denmark
DK-2800 Lyngby, Denmark

Jianbo Hu
Laboratory for Shock Wave and Detonation Physics Research
Institute of Fluid Physics
China Academy of Engineering Physics
Mianyang, Sichuan 621900, China

Hyotcherl Ihee
Department of Chemistry
Korea Advanced Institute of Science and Technology
Institute for Basic Science, South Korea
Daejeon, Republic of Korea

Mohammed Kaplan
Physical Biology Center for Ultrafast Science and Technology
Arthur Amos Noyes Laboratory of Chemical Physics
California Institute of Technology
Pasadena, California, USA

Sang Kyu Kim
Department of Chemistry
College of Natural Sciences
Korea Advanced Institute of Science and Technology
Daejeon, Republic of Korea

Amisha Kizhakkedathu
Aragen Bioscience Inc.
Morgan Hill, California, USA

Oh-Hoon Kwon
Department of Chemistry
Ulsan National Institute of Science and Technology
Ulsan 44919, Korea

I-Ren Lee
Department of Chemistry
National Taiwan Normal University
Taipei, Taiwan

Wenxi Liang
Wuhan National Laboratory for Optoelectronics
Huazhong University of Science and Technology
1037 Luoyu Road
Wuhan 430074, China

Christoph Lienau
Institute of Physics
Carl von Ossietzky University Oldenburg
26129 Oldenburg, Germany

Milo M. Lin
Green Center for Molecular
Computational, and Systems Biology
Department of Biophysics; Center for Alzheimer's
and Neurodegenerative Diseases
Pickens Biomedical Building, ND11.200
University of Texas Southwestern Medical Center
Dallas, TX 75235, USA

Charlie Qianli Liu
Vision-X Enterprise Management Ltd.
Shanghai, China

Lynne Martinez
Division of Chemistry and Chemical Engineering
California Institute of Technology
1200 E. California Blvd., MC 101-20
Pasadena, CA 91125, USA

Bengt Nordén
Chalmers University of Technology
Gothenburg, Sweden

Samir Kumar Pal
Department of Chemical Biological
and Macromolecular Sciences
S. N. Bose National Centre for Basic Sciences
Block JD, Sector III, Salt Lake
Kolkata 700106, India

Lawrence W. Peng
Campus Shared Services IT
Office of the Chancellor
University of California, Berkeley
Berkeley, California, USA

Jorge Peón
Institute of Chemistry
Universidad Nacional Autónoma de México
Mexico City, Mexico

Chong-Yu Ruan
Department of Physics and Astronomy
Michigan State University
East Lansing, MI 48824, USA

Mary Sexton
1458 Martingale Court
San Dimas, CA 91773, USA

Rajiv R. Shah
Innovation and Entrepreneurship
Jindal School of Management
University of Texas at Dallas
Richardson, TX 75080, USA

Dmitry Shorokhov
Physical Biology Center for Ultrafast Science and Technology
Arthur Amos Noyes Laboratory of Chemical Physics, M/C 127-72
California Institute of Technology
Pasadena, CA 91125, USA

Theis I. Sølling
Department of Chemistry
University of Copenhagen
Universitetsparken 5
DK-2100, Denmark

Jack A. Syage
Immunogen X
1600 Dove Street
Newport Beach, CA 92660, USA

William F. Tivol
Lawrence Berkeley National Lab
1 Cyclotron Road, Mail Stop Donner
Berkeley, CA 94720, USA

Giovanni Maria Vanacore
Institute of Physics
École Polytechnique Fédérale de Lausanne (EPFL)
CH-1015 Lausanne, Switzerland

Chaozhi Wan
Minioptic Technology
923 Coronado Drive
Arcadia, CA 91007, USA

Yuhong Wang
Department of Biology and Biochemistry
University of Houston
3455 Cullen Blvd
Houston, TX 77204, USA

Aiguo Wu
Ningbo Institute of Materials Technology
and Engineering (NIMTE)
Chinese Academy of Sciences (CAS)
Room A510, No. 1219 Zhongguan West Road
Ningbo 315201, China

Isaac de Melo Xavier, Jr.
Federal University of Pernambuco
Recife, Brazil

Tianbing Xia
Diagnostics Division
Abbott Laboratories
1921 Hurd Drive
Irving, Texas, USA

Ding-Shyue (Jerry) Yang
Department of Chemistry
University of Houston
Houston, TX 77204, USA

Nabeel Zewail
Factual Inc.
Los Angeles, California, USA

Maha Zewail-Foote
Department of Chemistry and Biochemistry
Southwestern University
Georgetown, TX 78626, USA

Dongping Zhong
Department of Physics
Department of Chemistry and Biochemistry
The Ohio State University
Columbus, Ohio, USA

Shouzhong Zou
Department of Chemistry
American University
Washington, DC, USA

1 Ahmed Zewail: Scientist, Humanitarian, and My Father

Maha Zewail-Foote*

I grew up on three major academic campuses. I was born at the hospital at the University of Pennsylvania, where my parents were graduate students. We then journeyed to Berkeley for my father's postdoctoral position in the laboratory of Charles Harris. Two years later, we moved to sunny Pasadena, California, where my dad joined the Caltech faculty. I remember running around the beautiful gardens of the Caltech campus, spying on the turtles and fish at the ponds, and listening to the legendary theoretical physicist, Richard Feynman, play the bongo drums. In the following years, I studied my father as he started as a junior faculty member at Caltech and grew into a world-renowned scientist. Undeniably, my father was brilliant. But, to me, what set him apart was his unwavering optimism, extraordinary perseverance, and continued curiosity for knowledge. Over this past year, after the death of my father, I have found myself holding on to all the meaningful memories and life lessons he offered me. I am sharing here a few of those precious and distinct memories that remain important to me. It is my hope that through my father's life story and his example, we can not only honor his legacy, but also be inspired by his dreams and seek a better future fostered by science and education.

* Email: zewailfm@southwestern.edu.

Growing up as the Daughter of "AZ"

When I was a young girl, my father would often take me with him to restaurants with one of his students or a collaborator. We usually ate at an Italian restaurant near the Caltech campus where my dad would order his usual — a pizza cooked well done. I would sit quietly and listen to the two of them talk about science. I was too young to understand what they were talking about or the impact their experiments would later have. But, I could always recognize the excitement and passion in their voices as they designed experiments, discussed the data, and generated new ideas. Those vibrant conversations would be recorded by jotting down their ideas on the backs of countless restaurant placemats and napkins. Memorably, I witnessed the scientific process in real time and the collaborative nature of academic science that would ultimately lead to significant breakthroughs. These were some of my early memories that ultimately sparked my own passion for chemistry and shaped my perspectives on science and research.

I became an undergraduate chemistry major at Caltech while my dad was in the midst of pioneering the nascent field of femtochemistry. It was an especially exciting time in his laboratory as his research group was examining a variety of molecular systems and reactions using ultrafast laser spectroscopy. During this period, I was a student in my dad's physical chemistry course, an upper-level course for undergraduate and graduate students that covered the latest in chemical dynamics. For each class meeting, my dad would enter the lecture hall wearing his coat and tie, which was his typical work attire. He was gifted in his ability to ignite students' passion for science. Part of this stemmed from his uncanny ability to simplify the most complex and sophisticated concepts by starting from the very basics and gently building our understanding step-by-step. As his student, I learned not only physical chemistry but also gained valuable insights that I would later use in my own teaching.

Later, when I was completing my doctoral studies in biochemistry at The University of Texas at Austin, I returned to Caltech for a week to visit my father and finish writing my dissertation. Each morning, we would stop at Starbucks on our way to the office to get an extra-hot cappuccino with his initials "AZ" written on the cups. I occupied a large conference table in my dad's office. Every bit of this space was entirely given to the purpose of being efficiently conducive to thoughtful work: the finest European chocolates were always nearby, and inspirational quotes and images of historical thinkers like Albert Einstein and Benjamin Franklin were hung on the walls. His office was meticulously organized with neatly stacked books and papers. This was also a place where my dad met with his students, so I had the opportunity to observe his mentoring style, something I appreciated as a graduate student who was interested in starting my own laboratory. In one particular meeting, my dad had ordered sandwiches for himself and his student while they were completing a manuscript for publication. The lunch lasted for hours as they went through the manuscript sentence by sentence, paying attention to every detail. Every word in the manuscript had a purpose. This was how my dad worked — meticulously and with great enthusiasm and passion. He enjoyed writing and conveying his ideas and tried to make each article a masterpiece.

As I worked on my own Ph.D. dissertation, my dad and I would often discuss my research. The project I was pursuing as a graduate student involved an anti-cancer drug and its interaction with DNA. In particular, I was evaluating the kinetics of the forward and reverse reactions, structural effects of drug binding, and the effects of these covalent adducts on DNA repair. My dad, being the lifelong learner, was fascinated by my research — particularly the kinetics — and our discussions seemed to always spark his curiosity. He would ask many questions to learn more about nucleic acid chemistry. As a student, I focused on the fine details of the experiments and the ensuing results while my father was always searching for the big picture and its far-reaching implications. These discussions were more than just two scientists discussing chemistry; they were a loving father and daughter being drawn even closer together through our common passion for chemistry and discovery.

Following in my father's footsteps, I became a chemistry professor at Southwestern University, a liberal arts and sciences institution in Texas. As I rose through the academic ranks to full professor, I deeply cherished our lively and invigorating conversations about science, academics, politics, and life. Amidst our busy schedules, we looked forward to spending quality time together during the annual Welch Foundation chemical conference and black-tie award banquet in Houston. My father served on the science advisory board for the Foundation while I was the Chair of my chemistry department. My dad and I would meet in Houston for several days every October. We would lounge by the hotel pool working side-by-side, usually on our respective manuscripts, and we would share stories, sip tea, and enjoy the ambience. I would always ask him about his newest scientific discovery, to which he would smile with his eyes sparkling and share with me his latest publication. I was always in awe of his innovative and creative work that pushed scientific boundaries, but I was even more impressed with his contagious joy for science, discovery, and learning. I respected his opinions and viewpoints and would seek his advice on whatever complexity or obstacle I was facing at the time. My dad would always offer a perspective that simplified the challenges at hand and walk us through a clear plan of action. Everything seemed so straightforward after our conversations and I always left Houston feeling empowered.

Our shared academic journeys eventually came full circle in special and memorable fashion. Back in 1994, as a Caltech faculty member, my father handed me my Caltech diploma at the commencement ceremony. Sixteen years later, in 2010, I handed my father an honorary degree from Southwestern University where he also delivered the commencement address. It was a singular, heartfelt moment of joy and pride for me as we stood on the stage together as father and daughter, both professors and scientists. During his commencement address, my dad summed up the "recipe of success in three words: passion, optimism, and opportunity." These are great words of aspirational advice for anyone, and they certainly expressed the principles of success by which I watched my father live his life.

These principles were inherent in my father's nature. Anyone who knew him well, or who worked with him for any length of time, could tell you that he was an eternal optimist. He did not dwell on problems, but rather searched for solutions, both in science and in daily life. This was an aspect of my father's personality that I admired greatly. His scientific breakthroughs to create the first ultrafast chemical camera stemmed from years of hard work. Along the way, a myriad of problems arose that all needed to be solved. But his path to great scientific achievement was driven by his optimism for the future, his attention to detail, and his relentless pursuit to move forward. In his scientific ventures, he kept his focus on his goals rather than the glare of seemingly insurmountable challenges. These characteristics that made him a phenomenal scientist also paved the way to his success in other areas. He used the same positive outlook and passion to make major humanitarian contributions to society and to inspire people from all over the globe.

Proud to be an Egyptian

My dad was always proud of being an Egyptian and cared deeply about his beloved native country. This pride was clearly evident in all aspects of his life. In his office, my dad listened to the legendary Egyptian vocalist Umm Kulthum while surrounded by Egyptian decor. For his laboratory, my dad chose to paint the walls a "Nile blue" color. And the personalized license plate of his easily recognizable car was "Musr," the Arabic word for Egypt. For movie night, my dad once selected a murder-mystery for us to watch together, which to my surprise was a documentary on the mystery of King Tutankhamen's death! Visiting friends and guests would be immediately captivated by a sense of Egyptian culture and would experience authentic Egyptian hospitality. For example, science would often be discussed over a cup of tea or coffee with gourmet chocolates or nuts. Everything was conducted in a "civilized" manner — a word my dad used often and embraced.

In his seminars and published papers, he would remind his audience about Egypt's long history of accomplishments, among them that Alexandria had been the intellectual hub of the world and was home to the first library and university as well as great scientists such as Euclid, Archimedes, and Hypatia. A long list of major scientific breakthroughs and achievements emanated from Alexandria, including the first proof that the earth was not flat. In addition to sharing stories of Egypt's rich history of leading the way in scientific discoveries and innovations, my dad would find ways to incorporate Egypt in his presentation images. For example, he depicted the branching of the femtochemistry field with the branches of the Nile Delta.

The Power of Knowledge

My dad deeply believed that the power of knowledge leads to the progress of society and worked tirelessly to bring cutting-edge research and innovation back to Egypt. He established the Zewail City of Science and Technology as a center of excellence and innovation and to help strengthen Egypt. My dad remained eternally optimistic about Egypt and its future and encouraged the new generation to pursue their dreams to become leaders, scientists, and entrepreneurs. He was appointed as the United States science envoy to the Middle East and served on President Obama's Council of Advisors on Science and Technology to help solve the economic and political challenges in the Middle East.

My dad also had a passion to advance society by promoting science and technology in Egypt as well as the Arab world and to widely communicate the importance of fundamental and curiosity-driven research. He traveled worldwide to deliver lectures about science and society. In his article, "Science for the Have-Nots" (published in 2001 in *Nature*, issue 6830 of volume 410, pages 741 to 747), he called for investment in education, resources to support research, as well as the creation of partnerships between developed and developing countries. Later, the two of us followed up on that article with the co-authored piece, "Science for the Haves" (published in 2013 in *Angewandte Chemie International Edition*, volume 52, pages 108 to 111), a work we were proud to call Zewail^2 (Zewail squared), where we discussed the changes in science over the past decades and the importance of encouraging discovery and innovation which stems from curiosity-driven research in order to achieve progress.

As I watched my father's life journey unfold as he became a world-class scientist, humanitarian, and scientist-diplomat extraordinaire, I witnessed a man who had a passionate curiosity about nature and a genuine thirst for knowledge. After winning the Nobel Prize in 1999, my dad continued to revolutionize science with his invention of 4D electron microscopy, a breakthrough technology that allows the visualization of structures at the atomic scale in *both* real time and real space. He genuinely enjoyed science and learning, and even after receiving the most prestigious honor in chemistry, it continued to be his true passion. One of his favorite places was the bookstore where he would acquire non-fictional books on science, economics, history, and biographies. These books, which were neatly arranged in his office, were sources of his inspiration and contributed to his lifelong learning.

Being the first Arab to win a Nobel Prize in science, my dad became a famous "rock star" scientist and was recognized all over the world. As he traveled the globe, he met with prime ministers and other world leaders in hope of inspiring positive change. With this notoriety, he remained a humble man who always remembered his journey from Egypt to the United States; from a young boy intrigued by the mysteries of the transformation of matter to a man who fulfilled his dreams to visualize the movements of atoms. Many people have shared their fond stories with me about how they met my father, heard one of his lectures, or how he inspired them to pursue their own dreams. When my dad met people or engaged with an audience, he was charming and charismatic. His warm smile, subtle sense of humor, and magnetic personality were captivating.

I am completing this essay drinking an extra-hot cappuccino and being meticulous in my description of my father — in true Zewail fashion. My father was a remarkable individual filled with hope, optimism, and perseverance. What my father instilled in me is what I always strive to pass on to my students and my own children: keep it simple, find your passion, work hard, and always be optimistic. I believe this quote from his 2011 Caltech commencement address says it all: "Always be guided by the light of your knowledge and wisdom to shape your future, the future of your country, and the future of the world." My dad's love for science and learning was contagious and he was a true inspiration. My father taught me by example and I am deeply fortunate to be one of his many students, a colleague, and a friend, but far more blessed to be his loving daughter.

Photo 1.1. Maha's high school graduation.

Photo 1.2. Graduating from Caltech. My father handed me my diploma.

Photo 1.3. An award banquet.

Photo 1.4. Our annual get-together before the Welch conference in Houston, Texas.

Photo 1.5. My dad with his grandsons.

Dr. Maha Zewail-Foote, the eldest child of Dr. Ahmed Zewail, is Professor and former Chair of the Chemistry and Biochemistry Department at Southwestern University. At an early age, she was fascinated with chemistry and shared the importance her father placed on education and the pursuit of knowledge. Following her passion for science, she earned a B.S. degree in chemistry with honors from Caltech and a Ph.D. in biochemistry from The University of Texas at Austin. She joined the faculty at Southwestern University in 2003 and was promoted to Associate Professor in 2009 and Professor in 2013. Dr. Zewail-Foote is passionate about improving undergraduate science education. She helped launch Southwestern University's Inquiry Initiative and has conducted seminars on implementing a flipped classroom in chemistry courses. The central goal of her research focuses on understanding the biochemical and biological consequences of drug-DNA interactions and elucidating the molecular mechanisms involved in DNA damage induced by small molecules. More recently, Dr. Zewail-Foote is examining the effects of DNA damage on non-B DNA structure formation and stability and the impact of DNA damage on DNA structure-induced mutagenesis.

2 Pushing the Frontiers

Nabeel Zewail*

I want to first thank Dongping for giving me the opportunity to share this reflection with everybody.

My dad was a big believer in ideas. He felt that they had meaning, depth, and were the reason one lived his or her life. He specifically believed in the ideas of science and learning, not only as purely academic pursuits (not to say that those are unimportant), but as something deeper. The constant pursuit of knowledge was dad's life calling and a path forward for humankind. My dad believed that enlightened people were better off economically and more at peace with one another. He also saw science as the tool which could bring countries and cultures together.

When I was a kid, my dad would occasionally take me to school and we had this saying between the two of us. He would say "push the frontiers" and I would respond "turn on your wheels." I'm not sure how exactly I came to that response (to be fair I was only six at the time), but he seemed to really like it. I even remember him printing it out on a blank sheet of paper, putting it in a folder, and showing it to me when I got home. It felt quite official that way. Ever since the early days of my youth, my dad loved that expression and it always meant something for the two of us and the rest of the family.

* Email: nzewail@gmail.com

Pushing the Frontiers

That phrase captured my dad's ethos better than anything else, scientifically and personally. I love the word "push" in that phrase because the frontier for him wasn't straightforward, apparent, or self-evident, but rather it was something that needed to be pursued and pushed.

When it came to pushing, I'm not sure I knew anyone who pushed harder. My dad loved to work and had the keenest of eyes for detail, but I'm not sure he would tell you that any of it was "work" in the way most of us would conceive of it. He was actually excited about his job. I remember when I got my first job and told him about what I planned to do with my vacation days, and I saw a look of mild disappointment in his eyes as if to say, "why are you excited about vacation?" My dad wasn't very good at taking vacations.

My dad always stressed the importance of the fundamentals in being able to creatively answer questions. He oftentimes credited his education in Egypt for equipping him with a strong grasp of scientific principles that enabled him to do much of his groundbreaking research. This approach unfortunately made him a liability in the "finish homework quickly and play video games" department. When I took advanced placement chemistry in high school, all of my friends thought the class would be a breeze for me, that I would just go home, give my dad my homework, and he would just do it for me (shows you how naive high school students are). Let's just say that is not what happened. If I had a question on problem number 45, to my dad that meant I didn't understand the first 44 and we had to go through the entire history of the atom before I got my answer. The importance of understanding the fundamentals and not rushing through anything was the lesson he tried to teach, which was lost on me then (and if he were here today, he would say still lost) but I just really wanted to finish my homework.

My dad also had a remarkable attention to detail. When I was taking geometry and was stuck on a problem, I went to him for help. He noticed that I had drawn a diagram imprecisely (the perpendicular lines were more acute). He took his black pen and scribbled "lousy" on the top of the page. At this point, I wasn't about to redo my homework, so I just went straight to school and as I was helping my friends with their homework, they noticed the "lousy" remark at the top and laughed endlessly. Luckily, my teacher didn't find it lousy and gave me a sticker.

To push the frontiers for anyone is a daunting task and to do so with my dad's background and circumstances is even more unlikely, but frankly I don't think my dad viewed himself as a real underdog. That mentality never manifested itself. I think that's because he was supremely confident in his abilities, but he also had a way of always viewing things positively. For most, being born on the Egyptian Delta in the mid-1940s would likely not be a typical advantage but my dad never viewed it that way. If anything, my dad viewed his story as a continuation of the story of his ancestors from the Pharaohs to the Arabs who pushed the frontiers of science. He even said in his Nobel speech that, if the prize was given thousands of years ago, the Egyptians would emerge as world leaders in science. My dad's optimism in his own abilities and circumstances, as well as his optimism in the Arab world and Egypt more specifically, drove much of his efforts later in his life.

My dad genuinely believed that the economic fortunes of nations in the modern world were directly correlated with their ability and willingness to educate their populations. He also believed that the example of East Asia could be followed in the Middle East under the right political circumstances. Through Zewail City, my dad was pushing the frontiers on what was possible in Egypt. Even when things weren't going his way and the project appeared to be too overwhelming, he always remained optimistic about it. That relentless optimism drove him to push the frontiers and in the summer of 2017, Zewail City will be graduating its first class of students.

My dad's constant work to push frontiers meant I didn't have a typical childhood. He was often busy with work and his big ideas. Traveling the world to speak on his latest paper or working on his newest venture, this meant that he wasn't there with us for a lot of things. Basketball games and debate tournaments were often missed. When he was there, he did make time to come to our soccer games, take us trick or treating, help us on our science fair projects, and teach us to ride a bike. Growing up, I never really gave where he was or what he was doing much thought because I didn't know better, but now as I reflect on his life and the lessons I take from it, I have learned about and realized what it truly means to be driven by something, by an idea, by that frontier, and to live your life with purpose, passion, and on your own terms.

I'm saddened knowing that he won't see me live out my dreams and goals, both professionally and personally, see me get married, or see my future children. More than that, I will miss our conversations and his unique way of viewing different situations. But I know that the lessons he taught me are for a lifetime and I know that his spirit lives on with us all.

I have yet to find the frontiers that I want to push, but if I've learned anything, I do know that to get there I will have to turn on my wheels.

Nabeel Zewail is currently a Data Engineer at Factual, a technology company helping to organize the world's location data. Before that, Nabeel worked as a Business Analyst at Acorns, a financial technology startup helping millennials invest through micro-investing. Born and raised in Southern California, Nabeel received his Bachelor's in International Economics from Georgetown University in 2015.

3 Zewail's Beginning at Caltech

Daniel R. Dawson*

I arrived at Caltech in the fall of 1975 as a first year graduate student in chemistry. My intent was to work with the well-known spectroscopist, Professor G. Wilse Robinson. My first year at Caltech Robinson was on sabbatical at the University of Melbourne, Australia. This was before the Internet, so my interactions with him were traditional pen and ink airmail letters, a painfully slow way to develop plans for one's future. The custom then was to give each first year student a desk in some research group as a way of bringing them into the department's culture. I was fortunate to be placed in the group of the then department Chairman, Dr. John Baldeschwieler. On a whim, I asked him if the department might have funds to allow me to visit Robinson in Melbourne. A few weeks later, I was on a plane to Australia. The day I arrived in Melbourne, Robinson informed me that day he had accepted an offer from Texas Tech University for a well-funded Welch Professorship. Robinson offered to take me along with him, with good support, but the prospect of trading Pasadena for Lubbock, Texas and Caltech for Texas Tech was not appealing.

I returned to Caltech unsure about my future and thinking hard about the other research groups in the Department. Baldeschwieler

* Graduate Research Assistant.
Email: danielrdawson@gmail.com

assured me the department would be hiring another spectroscopist, and soon. I believe the first time I met Ahmed Zewail was when he gave a departmental seminar as part of his interview process at Caltech. That would have been sometime in early 1976. I recall that I had a brief meeting with Ahmed as the department wanted him to know there were prospective students. Ahmed was hired shortly thereafter and returned to Caltech briefly to organize his new space in the sub-basement of Noyes. His initial research group consisted of myself, first-year graduate student Duane Smith, and second-year graduate student Tom Orlowski. Ahmed still had obligations at Berkeley, so he often came to Caltech briefly to meet with us and then return to Berkeley.

I remember that Ahmed had no grant support for research at Caltech yet. He had start-up money and support at Berkeley, but that was tied to his IBM fellowship there. He set Tom and me on track to observe and measure ultrafast coherent transients. He contacted equipment manufacturers with requests to test various items. In short order, we had a whole laboratory of electronics and other gear mostly on loan. The experimental set up was complicated and time consuming, and Tom and I worked many late nights. Ahmed would roll into town and work with us feverishly. I recall that he thought other groups were on a similar experimental track and he wanted to be first. He was also anxious to get his first Caltech publication out. I recall his elation when the experiment worked and we got decent photos of traces on the oscilloscope. The paper was published in the December 1976 issue of *Chemical Physics Letters*, a journal with a quick turnaround time from submission to publication. Ahmed wrote the paper incredibly quickly. It was his first publication from Caltech and came out before he was even there full time.

Working late into the night, Tom, Duane, I, and I believe an undergraduate named Roy Mead, would venture out exploring the steam tunnels that underlain the campus. One night, we found our way into the old synchrotron laboratory in one of the physics department basements. It appeared to have been forgotten and eventually all the equipment there would be discarded, so we hauled a few items that we could use back to our laboratories. These were things like rack mounts for electronics or old laboratory tables. When

Ahmed returned from Berkeley, he would ask us where we had gotten them from. Not wanting to get him involved in our questionable business, one of us responded, "Treasure Island!" which was the nickname we had given the locale. Ahmed then asked, "What's this Treasure Island?" and we responded that he did not want to know. After several back and forth exchanges like this, you could see a light come on and he laughed, "Oh, Treasure Island, this is good!"

Another anecdote that I recall had to do with the equipment we borrowed. We had some of it longer than the manufacturers expected. I am sure it was all eventually either paid for or returned, and I am also sure that new equipment was purchased because Tom was able to continue using that basic experimental set up. However, the manufacturers would call up demanding that their equipment be purchased or returned immediately. Ahmed instructed us to get the manufacturers to call him directly. He, in turn, would not take their calls. With this delay tactic, we were able to complete the experiment.

I do not recall having any formal group meetings during this period. Ahmed would meet us individually and suggest a huge raft of articles to read. It was clear from the breadth of his readings that he was thinking very broadly about how technological advances could play into our research. Both the technology and the experimental methods were of high interest to me. I had to push myself hard to understand the theoretical papers, and it was precisely these he would ask about when he saw me next.

Upon completion of this short project, Ahmed and I discussed my direction. We agreed that I would construct a picosecond tunable dye laser. Ippen and Shank of Bell Labs had just published their description of the first of such lasers and Ahmed believed this was the right tool for the laboratory. I started on the project alone but was eventually joined by the Postdoctoral Researcher Rajiv Shah. Although Ahmed spent relatively little time in the laboratory, he was always aware of the details and the progress of the project. He was not only up to date on all the literature but also in touch with a wide variety of scientists involved in related enterprises. In this way he learned about hardware advances. Whenever I wanted to make a particular purchase, he would make me justify that piece of hardware relative to other products. If you went to his office unprepared, you were brusquely dismissed with an admonition to not waste his time.

The ultimate challenge of this project was alignment. Rajiv and I invested hundreds of hours in an attempt to develop a systematic approach. During this period, Ahmed spent more time in the laboratory asking questions and making suggestions. I recall we had the instrument working and were characterizing the pulses by mid-1977.

I left Caltech and the Zewail group shortly thereafter for a variety of personal and professional reasons. As a result, I never got to know Ahmed well or personally. There was a clear cultural divide between the scholar from Egypt and the long-haired kid from Southern California, but there was no mistaking his drive or his intellect.

Dan Dawson grew up in the greater Los Angeles area in Southern California and attended the University of California, Santa Barbara (UCSB) majoring in physical chemistry. Dawson enrolled in the graduate program in chemical physics at Caltech in 1976. With a Master's degree in hand, Dawson took up a soft-money research position in the chemistry department back at UCSB. In late 1979, Dawson changed direction and took up a position as Director of the Sierra Nevada Aquatic Research Laboratory and Valentine Reserve, two units in the University of California's Natural Reserve System located near Mammoth Lakes, California. With this position, Dawson's focus switched to environmental chemistry, working on rain and snow chemistry and watershed nutrient dynamics. Dawson retired in 2016 after 37 years in this position and now lives in Ashland, Oregon.

4 The Early Years with Ahmed at Caltech

Rajiv R. Shah*

My first recollection of Ahmed goes back to mid-July 1976 when I was made aware of a letter he had sent to Bob Curl, a Professor in the Chemistry Department at Rice, stating that he was looking for a post-doc. I had just finished my Ph.D. in electrical engineering (applied physics) from Rice and I was starting to apply to a number of corporate research positions. Bob, Tom Rabson, my thesis advisor, and Frank Tittel all shared a large common space on the first floor of the Space Sciences building for their laser laboratories, where I had an office. Since I was already in the job market, albeit applying mainly to industry research laboratories, when Bob told me about this opening, I wrote in late July to Ahmed enquiring about the post-doctoral position and offered to visit him when I was in San Diego the following month for a conference. Ahmed called me up and invited me to meet him.

I went to San Diego in August 1976, presented my paper at the Society of Photo-optical Instrumentation Engineers Conference, and then stopped by Pasadena on the way to Berkeley to see Ahmed. The meeting went well and I went on my way to Berkeley and then returned to Houston a week later. Ahmed was at a conference and

* Postdoctoral Research Fellow.
Email: rajiv.shah@utdallas.edu

he sent me a handwritten note on 1 September 1976, sending me forms to fill out.

In the meantime, IBM San Jose Research Laboratories had invited me for an interview as well and arranged to fly me out from Houston to San Jose for a couple of days. While this was going on, I got a call from Ahmed on 20 September extending me a verbal offer to join his group as a postdoctoral fellow. His only condition was that I get back to him in the next couple of days with an answer regardless of anything that was going on with IBM, before he would send me an offer in writing. I conferred with Tom Rabson and his suggestion was that, since I was young enough, the door to industry opportunities would still be open to me in the future, so I ought to seriously consider Ahmed's offer. In addition, spending a year or two working for him would be a good experience.

I called Ahmed in a couple of days, as he had asked, and accepted his offer in writing on 22 September. I still went on my interview trip to IBM San Jose Research Laboratories and had a great interview over two days. I told them that I would like to reapply to them after my stint at Caltech.

The official Dr. Chaim Weizmann Postdoctoral Fellowship offer came to me in writing from Ahmed on 28 September as well as from President Harold Brown and Vice Provost C. J. Pings of Caltech on 14 October. I officially accepted the offer on 21 October and agreed to join by 1 November, thus beginning my two-year journey with Ahmed Zewail and Caltech.

I joined Ahmed and the Zewail group when I moved to Pasadena and began work on 1 November 1976. The group at the time consisted of Ahmed's first graduate students, Dan Dawson, Tom Orlowski, and Duane Smith. The group also included Ahmed's first undergraduate students, Kevin Jones and Roy Mead. Dan and I worked to set up the sub-picosecond laboratory, Tom set up the single mode spectroscopy laboratory, and Duane set up the optically detected magnetic resonance laboratory. For the most part, all three research efforts were just starting out. The latter two, however, were somewhat ahead at the time, having started a few months earlier. These two laboratories were set up with some newly purchased equipment as well as some old equipment obtained from elsewhere on campus. Roy provided support for the laboratory activities while Kevin supported the theoretical efforts.

When I joined in November, the sub-picosecond laboratory started out with an empty room. The vibration isolation table had just arrived, as had the Argon ion pump laser. Many components were starting to come in, which enabled us to start constructing the Rhoadmine-6G dye laser. Substantial effort was involved in designing mechanical components and the Plexiglas enclosure, which were fabricated by Bill Schuelke Sr., Bill Schuelke Jr., and Tony in the machine shop, as well as electrical components and equipment, which were obtained from Irv Moskovitz in the electronics shop. In early 1977, we accomplished the first lasing action from the dye laser. We soon moved on to getting the second part of the extended cavity built with the second jet stream for the saturable absorber and the acousto-optic modulator. The sub-picosecond laser was based on the Chuck Shank and Eric Ippen design from Bell Labs — the only such system at the time. We got that to work quite quickly and were able to see the mode-locked stream of pulses on an oscilloscope (see image). However, we ran into a few challenges. First, the system was very sensitive to alignment and needed a fair bit of "babying." Second, since the oscilloscope's time resolution was not high enough to tell us the duration of these pulses, we still needed to build a Michelson Interferometer and use a dry ice-cooled photomultiplier tube to plot an auto-correlation curve to determine the pulse width. For the first of these challenges, Chuck Shank met us when he came to Caltech to give a talk and confirmed that everything we were doing was indeed fine. The second challenge, which was to measure the width of the sub-picosecond pulses, took us a lot longer to solve. One of the other projects that got underway during early 1977 which I was very heavily involved in was the construction of an effusive molecular beam system to study coherent and incoherent transients in molecules at zero pressure, resulting in some of the first observations of these phenomena.

The Zewail group was a small, tight-knit bunch. Everyone was very young, for the most part uninvolved in romantic relationships, and was therefore able to spend a lot of time in the laboratory. Ahmed was the only one who was married, but he still spent long hours in the laboratory and office when he was not traveling to conferences, and he thus set the tone and expectation for others. The group not only worked together but also ate together very often. Lunches at The Athenaeum were common. Dinners at the Middle-Eastern restaurant on Lake Avenue near California Boulevard accompanied the long hours. There were occasional Friday evenings at The Athenaeum or pizza dinners that Ahmed hosted at Roma's Pizza on Wilson Avenue near Del Mar Boulevard. The most memorable were the occasions in the early hours of the morning long after midnight, when Ahmed would meet up with us at Marie Calendar's, IHOP, or Denny's for a midnight breakfast of pancakes and eggs, and discuss with everyone present there the details — progress and challenges — of the day's activities.

Within a year, the group began to grow with the addition of Sally Sheard, Barry Swartz, Michael Wert, Bill Lambert, David Millar, Albert Nichols, Joe Perry, and Sam Batchelder. As the group grew, it became harder to keep abreast of all the activities, especially as Ahmed began traveling more as well. Yet, despite having to travel to various conferences, Ahmed was very dedicated on staying abreast of all the efforts underway with his young and hardworking team. Phone calls to various members of his team late in the day (and sometimes at night) were common. With the burgeoning group size and increasing desire by members of the team to make significant progress with their research, some of the group dynamics and interactions began to evolve as well, to where the tightness and camaraderie of the group began to dissipate gradually!

I left Caltech and the Zewail group at the end of January 1979, when I joined the Semiconductor Research and Development Laboratories of Texas Instruments and moved to Dallas. Ahmed and I began corresponding again a month later. I saw Ahmed a year later in 1980 when I was visiting Caltech on a recruiting trip on behalf of Texas Instruments and we had dinner that evening at a restaurant on

Lake Avenue. Life then got busy for me on two-fronts — in the corporate world, and also my personal life. My last correspondence with Ahmed was in March 2000, after the announcement of his winning the Nobel Prize.

What I remember most about Ahmed is that he expected a lot from himself as well as others around him. He set highly ambitious targets and did all he could to achieve them. He was confident in his beliefs and his goals, which was accompanied by great insight and intuition. Most importantly, he also cared very much for those he knew and worked with, and was simply a great human being! Ahmed has been one of the most amazing people I have had the privilege of working with in my life and career.

Photo 4.1. Ahmed Zewail with Dan Dawson and Rajiv Shah operating the sub-picosecond dye laser in early 1977.

Dr. Rajiv R. Shah is a Professor at the Jindal School of Management at the University of Texas at Dallas, and is the Founder and Director of the Systems Engineering and Management (SEM) Program. After spending more than thirty years in various Fortune-100 high-tech companies, he also co-founded Timmaron Capital Advisors (www.timmaron.com), a chief executive officers and board of directors advisory firm, and The indusLotus Group, a technology and strategy management consulting firm.

Dr. Shah served as the Chief Technology Officer for Alcatel North America, and was Vice President of research and innovation and network strategy for four years. At MCI Worldcom before that, he led network architecture and modeling and participated in corporate-level due diligence for mergers and acquisitions. He was with Texas Instruments for seventeen years in various capacities including research and development.

From October 1976 to February 1979, Dr. Shah was at Caltech as a Dr. Chaim Weizmann postdoctoral research fellow in chemical physics. He has M.S. and Ph.D. degrees in electrical engineering (applied physics) from Rice University in 1974 and 1976, and a B.Sc. (physics, mathematics, and statistics) from Ferguson College, Pune University, where he was a National Science Talent Search Scholar of the Government of India. He also has an Executive MBA from Southern Methodist University in 1987. He has over 25 patents and over 50 publications.

In October 2014, Dr. Shah wrote a book with co-authors Zhijie Gao and Harini Mittal titled *Innovation, Entrepreneurship and the Economy in the US, China and India: Historical Perspectives and Future Trends,* which was published by Elsevier.

5 In the Shadow of a Voyage through Time

Jack A. Syage*

Principium

I'm sure that I'm not alone in eagerly signing up many months ago with a flood of memories and a desire to write something especially poignant about our time with Ahmed. Then reality hits and our busy lives collide with the deadline! However, what we are currently busy with can always wait another day as we don't often get the opportunity to put our imprint on history. Additionally, when I emailed Abderrazzak apologizing for having to withdraw, he responded with some encouragement and finished with a simple yet resonant appeal: "Please try." So here goes over a glass of bourbon!

I was fortunate to be able to attend the tribute to Ahmed at Caltech on 19 January 2017 and was overwhelmed with recollections of what were some of the best years of my life. I was thrilled to meet so many people that I knew from so long ago (let's just say they were contemporaries) as well as the later generation and the greater Caltech community that were all part of Ahmed's universe. It was overwhelmingly inspiring to be surrounded by so much brilliance.

I came into the group as a postdoctoral researcher in 1982 and it was truly an incredible time for me. There was not a single day when

* Postdoctoral Research Fellow.
Email: jsyage@immunogenx.com

I drove into the Noyes parking lot in the morning that I didn't think about how blessed I was to be working at this amazing university, in this amazing research group, with such amazing people, fulfilling my scientific dreams. We sometimes don't realize this until we leave, but I can truly say I felt this every day.

Now, this is not to say it was always a bed of roses and it shouldn't be. We can better appreciate the good things when we have some adversity as a frame of reference, so my story may reveal another side to what you are probably reading in this book.

We all know the jovial side of Ahmed. He was a true gem; a real enthusiast and it rubbed off on so many of us. There were few greater feelings than to get into a stimulating discussion with him, trading ideas, and challenging each other. Ahmed openly invited opinions and thoughts which was wonderful. There is nothing quite like the old-fashioned paper writing exercise when we would, at least during my time, use primitive word processors to draft a paper and then sit in Ahmed's office as he read and edited it in real-time (with a pen!) so that we could go back to the sub-basement and revise it. And should we not accept something or reword it, he would catch it! Of course we didn't need to do that too often because he was very intuitive and his edits were usually spot on.

The Doghouse

But there was another side where some of us would occasionally find ourselves in what was known as Ahmed's "doghouse." He was generally accepting of working hard on experiments that weren't straightforwardly panning out because such endeavors are indeed worthy and honorable challenges, and we all know Ahmed loved challenging scientific problems. He certainly wasn't going to pursue the easy stuff and none of us were there for that either (well, maybe some were!). When I joined the group in 1982, I was forewarned about a history of turmoil, but I was undaunted because I was hungry to do well and had faith in my capabilities.

One night at my desk in the sub-basement, when Ahmed was on travel, I took the opportunity to work on a paper that grew out of my graduate years. I had developed into an independent researcher during that time and was excited about a couple of theory papers that I was writing solo. All my material was fully spread out on the desk. Then I hear the elevator door open and the tell-tale sounds of hard-sole shoes walking down the hall toward my room. Now, graduate students and postdoctoral scholars wore nothing but sneakers or sandals, so my heart skipped a beat… it was Ahmed! I knew he wouldn't like me working on my own project, so in a panic I tried to decide what to do. Should I cut him off at the pass by getting up and meeting him in the hall or should I quickly gather up all my stuff and shove it into a drawer? Instead, I froze and faced the inevitable. He comes into the office and walks down to my desk with a big smile on his face wondering what I was up to. When he looked down, he became ashen. I can't recall what he said, but he quickly retreated with, to put it mildly, a displeased look on his face. I knew I was in trouble!

The next day when I arrived and settled into the laboratory, the phone rang and someone said that Ahmed would like to speak to me. I picked up the phone and Ahmed said to me in a very terse voice, "Jack, I'd like to see you in my office!" My heart started racing as I headed up to his office. This is how I remembered the conversation unfolding:

> Ahmed: "Sit down."
>
> Jack: "OK."
>
> Ahmed: "Jack, what does it mean to you to be working in this group?"
>
> Jack: "What do you mean?"
>
> Ahmed: "I don't think you appreciate the opportunity you have here."
>
> Jack: "Ahmed, I really do. I love it here and always think about how fortunate I am to be in this great group of yours." (And then I rattled on about the excitement of working in the group, amongst other desperate ramblings.)
>
> Ahmed: "How long have you been here?"
>
> Jack: "About a year and a half."
>
> Ahmed: "Maybe you should start thinking about where you should go next."
>
> Jack: "Ahmed…!"

Well, I thought I was going to get fired, which was especially distressing since my soon-to-be-wife Liz was just starting her own postdoctoral stint and I was hoping to stretch mine out a bit longer. The conversation went back and forth and he told me what was bothering him; that I probably did not value the projects I was working on in his group. I explained that I was working on a paper (I knew better than to say "during my free time") that was dear to me and could potentially allow me to further develop my mathematical skills, and that, at the same time, I'd also love to do a little more theory for him. Yes, there was some groveling! But it was more like a lover's quarrel and we both got teary. He was hurt and I was scared! As quarrels go, you work things out and eventually he started to soften and a smile started to show. I was beginning to feel a little less panicked, but I didn't really think I was out of the doghouse until he finally said:

> "Let's go get some pizza."

Place Your Bets

After my postdoctoral fellowship, I took a position at the Aerospace Corporation in El Segundo, California, in 1984, which is a government-supported but independent arm of the Air Force Space Program. I was fortunate to join the vibrant laboratory division and I was able to form a reasonably robust research group that kept me in the chemical physics game for many years. For a non-academic laboratory, I had satisfying success due to strong focus, reasonable funding, and a driven work ethic. Ahmed and I stayed in touch mostly through conferences, many of which were in Europe.

I was particularly pleased to have been invited to the 1996 Nobel Symposium in Chemistry held at Alfred Nobel's mansion in Björkborn, Sweden. This symposium is an annual, but little known, event that is truly a preview of which scientific fields will soon be honoured! An initial set of invitations were sent out to the usual luminaries in ultrafast spectroscopy, and those I knew were waving them around justifiably with pride. They were the invited speakers. Shockingly, a few weeks later, I got an invitation in the mail to attend (but not talk). I was clearly on the "B" list. But hey, only 40 in total were invited. My invite was owed to the fact that I was doing ultrafast on molecular clusters, which was a sufficiently novel field for me to be part of a small group of practitioners. Well, that symposium was clearly a Zewail Fest and signaled what was to come. It was an incredible time with someone actually impersonating Alfred Nobel, a historian totally acting in character; a real highlight! But Ahmed was undeniably and deservedly the star of the party. I know in this world you can bet on almost anything; if I knew where to place my bet I would have made out very well in 1999!

Group Member Forever

As I balanced my responsibilities of paying Air Force programs for launch vehicles and pursuing my independent research career (and also raising a family!), there was not much time for side ventures... with one exception! More than once, I would get a call from Ahmed and he would say, "Jack, there is this paper we need to write." A few times it was an invitation to write a review article. Although review articles can be notoriously time-consuming, how could I refuse Ahmed? It was not an easy task in the 1990s to look up and photocopy hundreds of references, especially at the Aerospace Corporation, but it was a great opportunity to highlight some of my own work as well as to be a co-author with a somewhat recognizable name (getting back at you Ahmed!). Getting instructed to do this and that, it was like I never left the group. Just recently, I was cleaning out one of my offices to move into another and rediscovered some of these old correspondences. Ahmed was a stubbornly analog guy. I would mail him a draft and a week later it would come back marked up with his copious handwritten edits and a type-written cover letter (clearly dictated to his secretary) thanking me profusely and informing me I needed to send him the update the next day! I'm sure I am not alone in these endeavours.

Operti

I started Syagen Technology in 1997, a mass spectrometry company focused on developing instrumentation for homeland security and pharmaceutical analysis, and faded from the chemical physics scene. But Ahmed and I did maintain occasional contact and my last encounter with him was when he won the 2011 Priestley Medal. I was at the American Chemical Society meeting in Anaheim, California, and spent many moments with him. This was during the period following the Arab Spring when he was spending a lot of time in Egypt working for constitutional reform and crusading with the Egyptian youth movement. There was a whisper campaign for him to consider the Presidency, which he vehemently and believably denied aspiring to. As we talked about this heady stuff, he quickly switched topics and asked, "So, how are you doing Jack, and how is Liz?" I was astonished that he remembered my wife's name 27 years after I left the group! Despite all the attention he was getting at the meeting and being escorted around (and his daughter Maha was delightfully part of the entourage), he genuinely found time for me. Later at the Priestley Award signing ceremony, I got in line and asked him to sign the book I owned, *Voyage Through Time: Walks of Life*

Photo 5.1. Priestley Award signing ceremony; *Voyage Through Time: Walks of Life to the Nobel Prize* tucked under Jack's arm. (Picture taken by Maha.)

to the Nobel Prize. He was taken with surprise and then he looked up at me with a smirk and asked, "So Jack, did you read it?" I was happy to be able to answer: "Every word!"

I didn't hear much more about Ahmed until his death. It really caught me, as I'm sure it also did for many of us, by surprise. It is a lesson to us all: Embrace life because even the apparent immortals eventually pass away. So, on to my next tumbler of bourbon to toast Ahmed and all he has done.

6 Meeting Ahmed Zewail

Isaac de Melo Xavier, Jr.*

When I was 12 years old, I started a chemical laboratory at home in Recife, Brazil. My little laboratory grew to have plenty of glassware and up to 200 reactants. At that time, I had a dream to study in an American university. I was accepted as an undergraduate student in the chemical engineering program at the Federal University of Pernambuco (UFPE) as it was the only chemistry course available at that time. From the chemistry courses I took, my favorite undergraduate textbook authors were John D. Roberts, Harry B. Gray, and Linus Pauling who, coincidentally, were all from the California Institute of Technology. In order to apply to the Brazilian agency Ph.D. fellowship, a Master's degree was required. Again, the only Master's program available at UFPE was in physics. I was accepted and worked under the supervision of Professor Rios Leite, who had just gotten his Ph.D. from Massachusetts Institute of Technology working with Michael S. Feld.

My experimental thesis was on the saturation absorption laser spectroscopy of carbon dioxide. In the physics department, Richard Feynman was a superstar, but usually you wouldn't hear about chemists. It was in the *Physics Today* issue of November 1980 about laser chemistry where I heard about Ahmed H. Zewail for the first time. Among several very interesting feature articles by Richard B. Bernstein,

*Graduate Research Assistant.
Email: isaac@de.ufpe.br

Richard N. Zare, and Yuan T. Lee, I was especially impressed with Ahmed's article titled "Laser Selective Chemistry — Is It Possible?" So I wrote to Ahmed describing my work on saturation absorption and my interest in laser selective chemistry while I was applying for graduate studies in chemistry at Caltech.

I was also accepted by Yale University, University of Pennsylvania, and University of Rochester, but I decided to join Caltech due to the gracious incentives that Ahmed gave me. He made arrangements for me and my wife Ana to stay at Athenaeum upon arrival and he also arranged for a Brazilian student named Marco Lima from Vincent McKoy's group to help us get settled in Pasadena, California.

My Caltech sub-basement life then started at 056 Laboratory, where "photon locking" (the optical analog of spin locking) was observed for the first time. I have fond memories of that time and I am grateful for the opportunity to be part of this outstanding group.

Photo 6.1. Ahmed's office in December 1995.

Isaac was born in Recife, State of Pernambuco, Brazil. He holds a B.E. in chemical engineering (Summa cum Laude) and a M.Sc. in physics from the Federal University of Pernambuco, Recife, Brazil. He holds a Ph.D. in chemical physics from the California Institute of Technology under the supervision of Ahmed H. Zewail and Aron Kuppermann. He had a postdoctoral research position at the University of Pennsylvania, working with Gregory A. Voth. He was Professor at Federal University of Pernambuco, Brazil, until his retirement.

7 Some Memories with Zewail

Fida F. Al-Adel*

How did I meet Professor Zewail? And what was the extent of our collaboration?

In the early 1980s, I was asked by our university, the King Fahd University of Petroleum and Minerals (KFUPM), to plan from scratch the establishment of a laser research center along with the acquisition of necessary support for the building and resources such as equipment and manpower, amongst other things.

The most prominent Arab laser specialist at that time was clearly Professor Zewail from Caltech. He was invited to visit KFUPM and sought for his advice and ideas. From the beginning of this assignment and before starting the project, I had wanted to learn more about supersonic jet spectroscopy so that it could be incorporated into my research agenda. I was then sent by KFUPM for one year to Professor Zewail's laboratory and my university accepted that the establishment of the center be postponed until I come back. I remember Professor Zewail coming to the airport to receive me with my family and he drove us to a nice hotel. Later on, with the help of William Pickering, the ex-director of California's Jet Propulsion Laboratory, we got a nice Caltech apartment. Everything went well; Professor Zewail was very busy but I could work with his students.

---

*Visiting Associate.
Email: ffadel@kfupm.edu.sa

On my return, I wrote a proposal for the Supersonic Jet Laboratory. After the proposal was reviewed and approved, I dedicated a large amount of my time to establishing the laser center. The greatest help that we got from Zewail afterwards was receiving his best postdoctoral scholar, Spencer Baskin, for about two to three years. Dr. Baskin certainly made his time count; he enabled the established Supersonic Jet Laboratory of the Laser Center to become functional and produced excellent papers. After he left our university, we continued to produce high quality papers in the field until we made a change in academic direction.

In one of his visits, Professor Zewail brought me one of the two gifts shown in the picture. A few months later, I invited Professor Robert Donovan from Edinburgh University to our laser laboratory, and to my big surprise he gifted me the second vase in the same picture. I was amazed because, among millions of possible symbolic gifts, both of them selected very similar ones. Professor Donovan was amazed too! Maybe both saw in me an admirer of French culture?

Photo 7.1. Right, a gift from Zewail to Al Adel; left, a similar gift from Donovan to Al Adel. The 2 gifts were made within a few month's interval without any coordination between Zewail and Donovan!

Dr. Fida F. Al-Adel was a professor of physics at KFUPM until 2016. He established a Laser Laboratory at KFUPM, where he is still doing research. He received the decoration of chevalier de l'Ordre des Palmes Academiques from France in 1994. His diplomatic experience spans from April 2001 to June 2005 as the Permanent Delegate of Saudi Arabia to UNESCO.

8 Memories of the Inimitable Ahmed Zewail

Spencer Baskin*

My association with Ahmed Zewail extended over the majority of both of our adult lives, so the reflections which are the subject of this tribute are broadly "delocalized" in time, to express it in the vernacular of quantum chemistry that Ahmed so much enjoyed using in everyday conversation. He played many roles in my life — teacher, supervisor, colleague, employer, staunchest critic and advocate, confidant, and friend — and, having known him for 32 years, with 26 of those years spent in a close working relationship as a member of his research group, there is little in my own biography in which he did not play a role. For this collection of reminiscences, I can only relate a few of the observations I made during those years that exemplify Ahmed's unforgettable and irreplaceable personality. These will be in the form of vignettes selected from decades of interactions that may appear random and unrelated, but in the best case, I hope they will provide insight into a remarkable man and what it was like to work with him. Although the purpose is not at all to recount or analyze Ahmed's well-known scientific achievements, most will naturally involve science as the context of the group's activities and illustrate essential characteristics of Ahmed the scientist. Of paramount importance to me, however, is providing a portrait of Ahmed the man.

* Graduate Research Assistant/Postdoc/Sr Scientist.
Email: baskin@caltech.edu

For this, let me begin, not at the beginning, but with a story which carries with it an emotional significance for me. The time was February 2011 when Ahmed had traveled to Cairo to exert whatever influence he could amid the chaotic events of the popular uprising against President Mubarak to guide Egypt onto a path toward a bright and peaceful future. His lifelong hopes for Egypt were in the balance, with the high drama and danger of revolution and rumors of his taking the role of president in a new government. Under these circumstances, it may be viewed as evidence of the most blatant naivety that I thought he would have any interest in or capacity to reflect on group business. Nevertheless I did, and wrote him an email on February 26 which began as follows:

> Dear Ahmed,
>
> I'll take the occasion of your birthday to convey my best wishes, and also update you on a few items of ongoing lab business. While your focus is naturally on other things of greater import, I'm sure you still would like to know that work is proceeding here. You can rest assured that most things seem to be in order, and no response is required, unless you see a need to give some input.
>
> First, my mother is doing much better now. …

The last was occasioned by the fact that Ahmed knew my mother was having some serious health concerns which required me to spend time with her in Texas. The rest of my message consisted of reports on a variety of papers in progress, group meetings, equipment repairs, and purchases. I never knew whether he actually gave any thought to the points of group business, but the following reply arrived within a few hours, in the middle of the night, Cairo time:

> Dear Spencer,
>
> Many thanks. I am glad your mother is doing well; give her my best wishes.
>
> I look forward to being back soon.
>
> Best regards,
> Ahmed

Photo 8.1. At the Zewail home. (Photo by Qian Baskin.)

It brings a lump to my throat to read this now as I reflect on this extraordinary expression, beyond any expectation, of his heartfelt personal concern, exemplifying the natural human touch that I was privileged to experience and that endeared him to many others.

Working with Ahmed over so many years has resulted in the two of us appearing together in a large number of group photos, many of which are included in this book. For a more personal photo, I have chosen the one shown in Photo 8.1, taken by my wife when my daughter, Bernice, was six weeks old, which shows Ahmed in his role as the abovementioned warm and benevolent family friend.

To set the stage for more stories of my life with Ahmed, a little background is appropriate. Also, in order to avoid the need to repeatedly qualify these stories as "to the best of my recollection," I will emphasize here that I will be conveying what are in fact sometimes distant memories and impressions, and apologize in advance for any inaccuracies that may unintentionally but inevitably creep in.

My first exposure to Ahmed and his research came from discussions with David Semmes whom I met when we were both first-year

Caltech graduate students living in Braun Dorm. David was a chemist and already a member of the Zewail group when I began seeking a research advisor after finishing one year of the course work required in my applied physics program. Based on David's description of the work and given that I had a longstanding interest in lasers and optics (I ordered a brochure on lasers from the U.S. Atomic Energy Commission in Oak Ridge, Tennessee around 1969, and I still have it), I went to see Ahmed to enquire about the possibility of joining his group. That first meeting went well — he needed another graduate student apparently, and I was very happy with the enthusiastic welcome that I received. I went to work fairly soon thereafter with his group of twelve research members in the Noyes sub-basement, sharing an office and laboratory 036 with David and Peter Felker. Such was the casual beginning that set the stage for my entire career and put me on a path to be carried along on a stimulating, rewarding, and highly unanticipated journey over the following decades in the Zewail group.

While my first experience with Ahmed offers a routine example of him exercising his warmth and charm on a prospective graduate student, a more remarkable indication of his personal charm can be found in a story that Tina Wood used to tell about Ahmed's interview trip to Caltech in 1976. Tina was working as a secretary in Noyes at the time and only saw Ahmed during his visit as he navigated his way around Noyes for meetings with various faculty members. Nevertheless, he impressed her so positively during this limited interaction that she said she was convinced that he had the potential to be a great asset to Caltech. Ahmed mentioned in his autobiography that she lobbied for him to be hired, but as she told me her version of the story, she also immediately let it be known that when he was hired, she wanted to be assigned as his secretary. Both of these wishes were ultimately fulfilled, and during the many years she worked for Ahmed, she even traveled to Egypt to help him with organizing the International Conference on Photochemistry and Photobiology in 1983. Sadly, Tina passed away shortly after Ahmed, so we are no longer able to hear any of these stories directly from her.

In my early experience in research in the 1980s, I was exposed, as I believe were most of his collaborators, to a number of Ahmed's scientific principles. One that I heard him repeat often was his disinclination to trust results from a computer program without convincing direct verification of their validity. From personal experience, this principle of his was exemplified when I was asked to confirm the rotational constants of the stilbene molecule (based on published bond lengths and bond angles derived from gas-phase electron diffraction measurements) by manual hand calculation, no computers involved. In line with this wariness of the danger of reliance on "black box" computations, he placed great value on "back-of-the-envelope" calculations that could capture the essence of a physical problem in a transparent fashion, and he frequently carried out such calculations by hand on a pad of paper, demonstrating in the process his impressive knowledge of basic physical equations and the values of the physical constants relevant to the problem.

This exercise of care in theoretical treatments had its counterpart in the conduct of experiments as he urged extreme caution against being misled by experimental artifacts. I heard him mention numerous times an early observation of "quantum beats" in fluorescence which was ultimately found to arise from electrical ringing in the signal of a photomultiplier. He was therefore always thorough in outlining control experiments or "checks" to eliminate any possibility of reporting erroneous observations. That he held it to be of the greatest importance to provide accurate measurements remains to me a true indication that he was an experimental scientist at heart. He would say that over time, interpretations and theories may evolve, but solid experimental data would never lose their value.

A mild example of Ahmed's rigor in experiments and his style of interaction with group members is also connected with stilbene (I suspect stilbene is the most-studied molecule in the entire history of the Zewail group). Our first dedicated search for rotational coherence in the supersonic molecular beam of 036 was carried out on the *trans*-stilbene molecule. The rotational recurrence period (based on the aforementioned rotational constant), as previously calculated by Peter Felker, was expected to be about 1980 ps, at which interval a dip in the fluorescence signal was predicted for detection of perpendicularly polarized fluorescence from cold, isolated molecules coherently excited by ultrashort pulses. The laboratory set up by Bill Lambert and Peter was perfectly suited to this study, with a picosecond, synchronously pumped, mode-locked, and broadly-tunable dye laser, a supersonic molecular beam, and a newly purchased microchannel plate fluorescence detector capable of 40 ps temporal resolution by means of time-correlated single-photon counting.

The first successful recording of the rotational coherence signal that we were seeking, initially seen on the screen of the multichannel analyzer in 036 by Marcus Dantus and me when we returned to the laboratory after lunch one day, was of stilbene seeded in a helium carrier-gas expansion at 45 psi. I saved the file, fit it, and plotted the transient with the fit residual (Photo 8.2) on our pen plotter. Then, after no doubt first showing it to Peter, I very confidently took the plot to Ahmed's office to show him that we had indeed observed

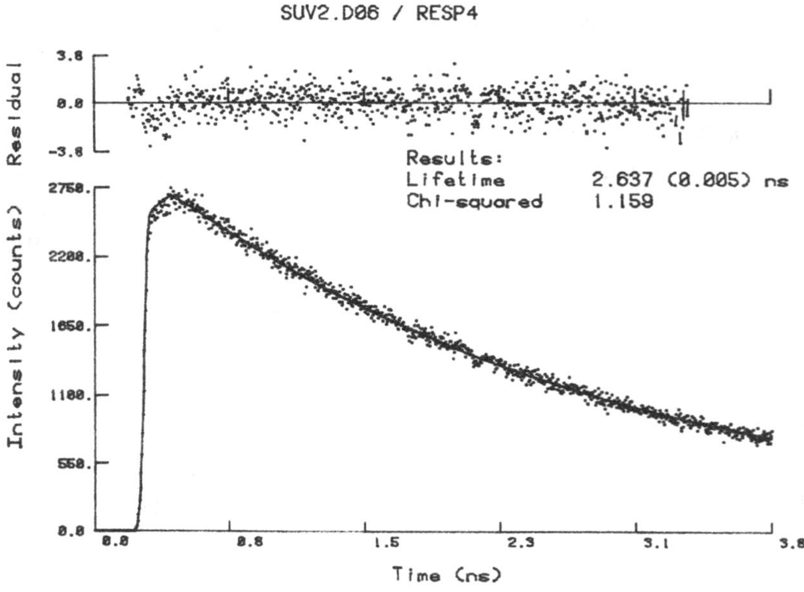

Photo 8.2. Polarized fluorescence transient of jet-cooled *t*-stilbene, recorded 6 December 1986.

the dip. He took it and, after carefully examining it from all angles, asked, "What dip?" Though I insisted that it was there, Ahmed was not yet ready to offer his congratulations, saying words to the effect of: "Nobody can see anything there without a microscope!" He thereby made it clear that he felt the need for more convincing results first, and this incident led him for a period of time to refer jokingly to my having "microscopic eyes."

This particular example illustrates Ahmed's standard routine of playing the devil's advocate to help us build a case that was unassailable by skeptical reviewers. We followed up our first recording with measurements changing the carrier gas and pressure (to lower the temperature in the molecular beam), as well as with detection of parallel and magic-angle polarization, all with the predicted results. Finally, the test that Ahmed cited as most conclusive was the measurement of deuterated stilbene and the stilbene-argon van der Waals complex, in which the observed recurrence periods tracked the calculated changes in rotational constant, thus providing a temporal "finger print" to identify the source of the fluorescence. The required experiments were all done in 10 days, a mercifully short time to reach the "Zewail threshold" for publication.

One consequence of the high bar that was generally set by Ahmed has been illuminated for me in recent days as the pending closing of our laboratories has caused concern among group members waiting for referee responses to submitted papers. Several have indicated their belief from experiences elsewhere that there was a good chance for a referee to insist that more experiments be conducted as a condition for publication. This possibility seemed very remote to me since it never occurred in my memory, but I then realized that Ahmed was likely to have already insisted that every imaginable control experiment be run before we submitted our papers in the first place. These he would be careful to report prominently, often in a list, in the manuscript to leave no room for doubt or misinterpretation.

Another memorable aspect of Ahmed's personality that I observed frequently over the years was his deep appreciation of American tradition and history. While devoted to Egypt, his admiration for the dynamic character and great achievements of his adopted country was evident, and these traits had obviously and fundamentally stamped their influence on the course of his life. In the pantheon of American pioneers in science and technology, he felt a special kinship with Benjamin Franklin, starting at least from his years of residence in Philadelphia (more on Franklin later). It became a tradition to have a group dinner each year during the week of the quintessential American holiday of Thanksgiving, and a discussion of the cultural and political history of Thanksgiving formed a regular part of the dinner conversation. In later years, there were often not many native Americans in the group, so Ahmed typically called on me to relate the history as a part of the cultural education he felt was important for those new to the country, as he had once been.

He never tired of absorbing stories of how Americans had transformed the world through extraordinary innovations. On a visit to Pasadena by my brother and sister-in-law in the fall of 2015, we stopped by Ahmed's office to say hello, and the conversation eventually turned to the biography Ahmed was reading on the story of the Wright Brothers written by David McCullough, which had just been published that year. Ahmed expounded enthusiastically on the incredible tale of the brothers from Ohio, whose determined and independent quest to achieve powered flight in competition with the well-funded scientific establishment represented by Alexander Graham Bell and the Smithsonian Foundation was, in his opinion, representative of the spirit of American ingenuity at its finest. His wholehearted recommendation of the book, with the same bubbling enthusiasm that I imagine had charmed Tina some 40 years earlier, made such an impression that I felt it would make the perfect gift for my brother as a Christmas present. I could not resist a chuckle when I opened my present from him on Christmas day that year to discover that, under the influence of Ahmed, we had each given the other a copy of McCullough's *The Wright Brothers*!

Speaking of my brother brings to mind the warmth with which Ahmed always received my family in their visits to California. He had met my parents first, and when he received the Welch Award in Chemistry in October 1997 from the Welch Foundation at a banquet in Houston, Texas, he invited my mother and me to attend as his guests. Two years later, in mid-1999, Ahmed accepted an invitation to my hometown of Waco to give the 2000 Gooch-Stevens lecture for the department of chemistry and biochemistry of Baylor University, giving my parents and my brothers (as well as their families) the chance to host Ahmed on the occasion of his visit. The Gooch-Stevens lecture series was distinguished by the fact that the majority of its previous speakers were Nobel Prize winners, and as fate would have it, Ahmed was awarded his own Nobel Prize between the invitation and the date of the lecture in early 2000.

The subject of the Nobel Prize evokes another distinct memory that I have of Ahmed. I first heard Ahmed's name and the Nobel Prize mentioned together when I rejoined the Zewail group in 1995. The password on at least one of the group computers was "femto95," and when I asked what the "95" meant, Soren Pedersen told me that it referred to the expectation that Ahmed would win the Nobel Prize that year. While considering this idea highly speculative at that time (obviously in my ignorance!), I correlated it thereafter with the habit that Ahmed noticeably had of being away from Pasadena annually at the time of the Nobel announcements in the first weeks of October. This was so regular an occurrence that the fact that he had made no such plans in 1999 struck me as significant enough to pay particular attention to the announcement of the Chemistry Prize that year, and I was astounded to hear his name announced as the winner. I later asked him if he had *known* something in advance, and his answer was naturally no, but I failed to ask the more pertinent question of what he had *expected*.

No description of Ahmed can be considered complete without devoting attention to his rather distinctive sense of humor. I imagine that one of the factors which solidified my role in the group was being the brunt of many of the jokes he enjoyed making as an intrinsic part of the group ambience. One example was his regular quip when I would go upstairs to see him: "Oh no, Spencer's here. There must be some problem in the lab. I wonder what it's going to cost me this time?" There were also his unfailing references to my "Texan English" (not meant as praise!) whenever we were struggling to find

Photo 8.3. Amusing demonstration to a student in the group conference room, November 23, 2010 (photo courtesy of I-Ren Lee).

the right way to clearly express some difficult idea in a paper we were writing. The good-natured intent of his remarks was always made clear by the twinkle in his eye and mischievous smile that accompanied them. A fine example of this look, with me clearly as the target of whatever joke was in progress, is captured in Photo 8.3, taken at I-Ren Lee's thesis defense celebration in 2010.

Group retreats gave Ahmed the stage to indulge in storytelling and jokes, and one of the latter that I remember has a special meaning these days. It went like this: A scientist died before he could publish the results of an important research project to which he had been very devoted. When he arrived in Heaven he was asked if he had any requests. He said that it was his great desire to fulfill his life's work by publishing his last results. Since this was a reasonable request, he was granted permission to write and submit his work for publication. He happily and eagerly went about preparing a manuscript, and after sending it off to the local science journal, he awaited a response.

He waited and waited and the time dragged into months without a reply. His frustration finally grew to a point where he called the journal and asked the editor why it was taking so long. The editor was apologetic and explained, "We're sorry, but we're doing the best we can. You have to understand that, in Heaven, the reviewing process is a real bottleneck, because *virtually all referees are in the other place.*"

To tie this angelic (or devilish) humor into Ahmed's world view, I must say that I saw his love of science as a way to reveal the wonders of nature without ever feeling that existence was reducible to a purely materialistic description. I never had theological discussions with him, but he was not reticent to express his convictions of the reality of a spiritual dimension to life, of which his scientific inquiries only served to increase his deep sense of awe. Because of this, I'd like to conclude with a quotation that Ahmed had framed on his office wall which came from another renowned scientist that he greatly admired, and with whom his career had several points of intersection. It is the self-written epitaph of Benjamin Franklin, given here in the same format as on Ahmed's wall, though without the beautiful calligraphic flourishes of the artist (James L. Wood, Tina's husband):

The Body
of
Benjamin Franklin, Printer
(Like the cover of an old book,
Its contents torn out,
And stripped of its lettering and gilding)
Lies here food for worms.
Yet the work itself shall not be lost,
For it will (as he believes) appear once
more
In a new
And more beautiful Edition
Corrected and amended
by
The Author

I read these words with difficulty at Ahmed's memorial service in Los Angeles last August, for they describe clearly the irrevocable loss experienced by those close to him as we continue on with our lives, but I believe that they also express his sense of a higher plane of existence and hope of a better destination after a life well lived. And yours was truly an extraordinary, full, and consequential life by any measure. We miss you, Ahmed!

Spencer Baskin, a native of Waco, Texas, earned B.S. and M.S. degrees in physics from Georgia Tech, studied at the Eidgenössische Technische Hochschule in Zurich, Switzerland, and taught high school math and physics at the Institut Kizito in Isiro, Zaire (Democratic Republic of Congo), before earning a Ph.D. in applied physics from Caltech under the direction of Professor Ahmed Zewail. Following five years of research split between the Research Institute of the King Fahd University of Petroleum and Minerals in Dhahran, Saudi Arabia and the chemistry department of the University of Houston, he returned to Caltech where he has spent 22 years working in nine different laboratories as a senior member of the Zewail group. For the last 10 years, he has been active in the second generation laboratory for Ultrafast Electron Microscopy which he joined at its inception.

9 Ahmed Zewail Memory Tribute

Lawrence W. Peng*

How I got to Caltech

It is hard to believe that it has been a generation since I graduated from Caltech under Ahmed's guidance. As a young teenager, I became aware of Caltech via public television and textbooks, and learned of scholars such as Feynman, Pauling, Millikan, Morgan, Richter, and Von Kármán, amongst many others. Being an impressionable kid, I decided I had to try and get into Caltech for college. That did not happen for college, but to my joy it did happen for graduate school.

So I got accepted by Caltech in early 1984 and prepared to make the trip west in August 1984. But I did not know what to expect, which Professor I would work for, or what research problem was of interest to me. It certainly would not be in synthetic chemistry as that has never been my forte. However, I knew it would be something involving kinetics, dynamics, optics, and mathematics. Fortunately, I had two current contacts with connections to Caltech's Division of Chemistry and Chemical Engineering which gave me some clues.

First was Andy Axup, a Purdue classmate who was already working in Professor Harry Gray's group. Of course, working with Harry meant he had to issue the challenge to take Harry's "Advanced

* Graduate Research Assistant.
Email: peng2@berkeley.edu

Ligand Field Theory" class (Chemistry 213a). So naturally, I had to take him up on that challenge. Today, I can proudly say I am one of the few who has taken not only 213a but also 213b-c.

The second was Dr. Duane Smith, who had recently become an Assistant Professor of Chemistry at Purdue. It turned out that Duane was Ahmed's first graduate student and he enthusiastically informed me about what it was like working with Ahmed and being part of his group. That discussion with Duane was where I was first clued in to Ahmed's intensity and scientific enthusiasm. I recall Duane in effect saying that you will never work harder in your life under Ahmed, and that turned out to be pretty much spot on.

So I arrived on campus shortly before classes began in the fall of 1984 and stayed at Marks Graduate House that first year. The first thing that surprised me was how small the campus was relative to Purdue or University of California, Berkeley. The second was how much more culturally diverse Southern California is compared to Missouri, Indiana, or even upstate New York. In any event, I did the initial exams, got my first teaching assistantship assignment (supervising the chemistry freshmen at Mead Laboratory), and heard the synopses from the chemistry faculty about their research programs. It was, on the one hand, an intense time trying to digest the topics the professors were studying, and on the other hand, also a reminder of how much more there always is to learn.

By that time, I had decided that I wanted to work in Ahmed's group. Through Tina Wood, his then executive assistant, I arranged for a time to meet Ahmed on a Friday afternoon after finishing my teaching assistantship work at Mead Laboratory.

This first meeting was one that I will never forget, and I suspect Ahmed never forgot either. I set out for the meeting after completing my teaching assistantship work, but the head of Mead Laboratory, Jane Raymond, also needed to ask Ahmed a question and requested to come along with me. Who am I to say no? Jane had her friendly German shepherd dog with her as we headed to Noyes Laboratory. We arrived at Ahmed's office and when he opened the door, the dog happily pounced on him, startling poor Ahmed! This was not the way I ever imagined my interview with a potential advisor to be; my name was from then on associated with a lunging dog! But fortunately, Ahmed did not hold that against me and after a quick chat, I was accepted as a member of the group. I found out later that I was lucky as Ahmed only accepted one new graduate student that year.

Immediately after that initial interview, I had a follow-up meeting with Ahmed and fellow graduate student Brian Keelan. Ahmed told me that we would discuss my experimental results the following Monday! This gave me a taste of Ahmed's quick scheduling. Fortunately afterward, Brian calmed me down by telling me that he and I would work together and present the results to Ahmed on Monday. With that, I began my graduate student career with Ahmed.

Initial Breaking into the Group

Over the course of my tenure with the group, I would always focus on trying to do my work the best I could while sometimes keeping Ahmed at a distance. I suppose it might be because I was initially a bit intimidated, or later to simply avoid lectures that I was not ready to hear. Ahmed and I got along pretty well, so in retrospect I should have tried to be more interactive with him.

I did allow myself a few non-work distractions. I played on a chemistry softball team named "Pitch and Bitch" as well as a volleyball team whose members were mostly from the Zewail, Kupperman, McKoy, and Weitekamp groups. I always went to one Dodgers' home baseball game when the Cardinals were in town and hit the movie theatre once per academic term. But my life was otherwise mostly reading, studying, and working. Although I later moved to an apartment two blocks from the parade route, I never saw the Rose Parade in person until realizing that 1990 was my last chance. Likewise, a visit to Disneyland for the first time in a generation just did not occur to me until late in the summer of 1990.

Initially, I spent my time divided between coursework and learning about the 046 laboratory in detail. The 046 laboratory might be called the "slow lab" since we were using nanosecond lasers, but there we did spectroscopies of anthracene, stilbene, and (dimethylamino)benzonitrile to provide "static" data (energy levels, vibrational levels, and rotational band structure) that the ultrafast picosecond and later femtosecond systems could use to help determine optimal wavelengths for their experiments. 046 was originally built by Brian Keelan, Jack Syage (a postdoctoral scholar from Brown), and postdoctoral scholar John Shepanski. I would be inheriting 046 once Brian graduated, so Ahmed wanted me to know the laboratory inside out.

Over the next two years, I learned about high vacuum, pulsed valves and supersonic jets, lasers, electronics, and the chemistry machine shop (operated by Tony Stark, Guy Duremburg, Delmer Dill, and Ray Garcia). The high-end laboratory computers at that time were a Digital PDP-11 in the picosecond laser laboratory in Room 036 and the Digital VAX system at Crellin laboratory. We used five-inch magnetic floppies and rolls of magnetic tape to store data along with strip chart recorders and ink plotters to create figures.

This was where I began to get into the habit of taking detailed notes for any lab I worked in. I noted every single setting, location, cable coding, connection, workflow, tweak, color, and items were deliberately raised off the floor. Brian related to me the earlier famous flood of the Noyes sub-basement after the sump-pump in the mechanical room across from 036 failed. A loud 10 Hz piston pump in 046 was especially problematic since the 046 laboratory laser system pulsed at 10 Hz. I got tuned into safety procedures (chemical, optical, and electrical), especially after getting a nasty electrical shock from the discharge of the Nd:YAG flash lamps. I was otherwise only in contact with wood so I had a partially numb hand for a day, but high voltage electrical safety has been on my mind ever since. The shock pushed me back against the wall between 046 and 048, and my fist punched a hole through the wallboard. The hole was put to good use when I had to disassemble, rotate, and reassemble 046 to support experiments in 048 a couple of years later. As it turned out, the hole was exactly where it needed to be to join the 046 laser output to the 048 experiments!

I got into Ahmed's doghouse for the first and only time during the re-orientation of 046 to support 048. The re-orientation required drilling holes into a supporting concrete wall in order to move a bunch of shelves which housed equipment, so I arranged for facilities to come in and bore holes into the concrete. That meant that after almost a week, I had not made much visible progress with re-orienting the lab. While getting ready to head out to Arcadia Valve and Fitting to get parts one day, I got Ahmed's call to "drive with me." So I got into his car to get the parts and along the way I got the lecture. The message was received, and a couple of days later everything was finished.

The work in 046 got me through candidacy exams, but by that time the group was already pushing towards femtosecond laser technology. This would lead to the construction of the group's first femtosecond laboratory in Room 047 (later called Femtoland 1 if I recall correctly), and shortly thereafter another laboratory in Room 047A.

Owning Automation

One of the experimental goals of the laboratory in 047A was to study the dynamics of small clusters with the hope of detecting picosecond spectroscopic changes as cluster size increases. It was built by Jack Breen, a postdoctoral scholar from Penn State, graduate students Dean Willberg and Peijun Cong, and myself. 047A employed a pump-probe design with picosecond lasers, supersonic jets, time-of-flight mass spectrometry, and associated control and signal detection equipment. As was typical of the time, there was a mixture of hand-built items and equipment bought off-the-shelf. The 047A laser system can be seen in the 1987 Zewail group picture shown in Photo 9.1.

The four of us drew straws to determine who had principal responsibility for particular aspects of the laboratory. The laser system went to Dean, the vacuum system including the time-of-flight spectrometer went to Jack, the instrumental electronics went to Peijun, and the laboratory automation, integration, computer control, and data analyses went to me. Initially, I thought I got the short straw. We were all chemistry graduate students or postdoctoral scholars, and computer control was simply regarded as a necessary side project. After all, the thermocouples and pulsed valve system controllers in 046 were manually programmed in RPN, and the code was printed on thermal paper which was scotch-taped to a laboratory book if it was necessary to retype the program.

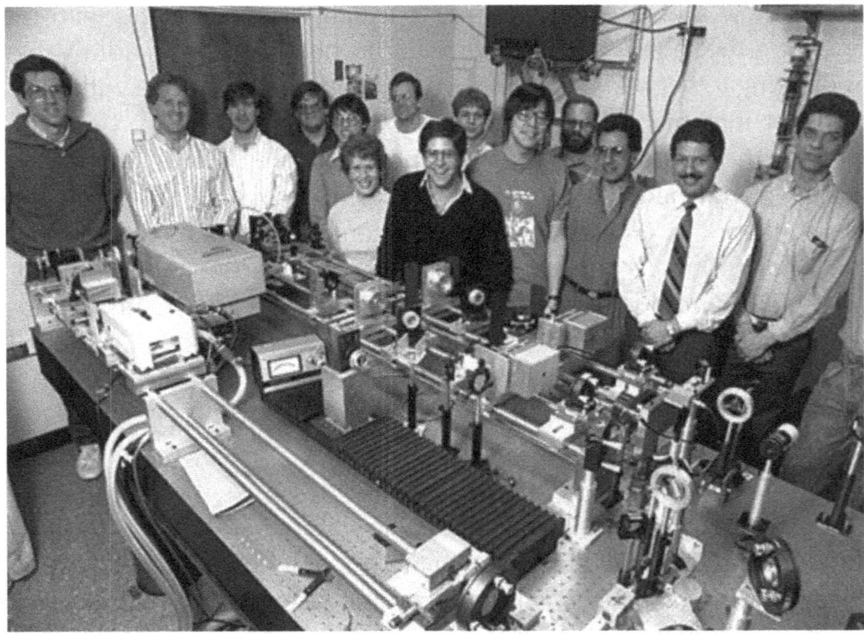

Photo 9.1. The Zewail group in 1987. From left to right: Jack Breen, Mark Rosker, Bob Bowman, Dean Willberg, Peijun Cong, Tina Wood, Earl Potter, Marcos Dantus, Paul Tripodi, Lawrence Peng, Spencer Baskin, Ahmed Heikal, Ahmed Zewail, David Semmes. (Photo taken in Zewail laboratory 047A.)

Beginning with Femtoland 1, we started moving to micro-computers with the IBM PC 286, which just displayed a bunch of numbers and still used 5.25 inch magnetic floppies. The migration to microcomputer control was complete with the introduction of the Macintosh II and the IBM 386 systems in 1986. The Macintosh versus PC versus UNIX wars were in full force by then.

At that time, the Macintosh won in light of a few important factors. Although the IBM PC (and nascent clones) had an initial cost advantage, the Macintosh had an intuitive graphical user interface (GUI) which allowed for easier visualizations and could show real-time data to research funding sponsors. The Macintosh and LaserWriter being the center of the new desktop publishing field would be a long-term boom to reducing publication preparation costs. The Macintosh plug-in card architecture was superior and higher performing relative to other platforms. Sheer computing power was the realm of UNIX workstations, but GUI systems were clearly the future despite their steep programming learning curve and continuous monitoring. Thus compared to PCs, the Macintosh had more computational power, ability to use more RAM, and the GUI heavily tipped the balance in its favor.

The Macintosh platform was chosen around the same time we started to build 047A. Femtoland 1 (047) was also upgraded to the Macintosh II. Some early Macintosh prototype software was brought to Caltech by Mark Rosker who was a postdoctoral researcher in the Zewail group from Cornell. The initial software was limited, but it could receive data from a detector, send simple RS422/232 commands to a stepper motor, plot the data graphically, and save the data to a file. A third-party Macintosh application called CricketGraph was used for creating plots which we would then send for publication.

For my laboratory 047A, I wanted to see a better integrated system suite which included data acquisition and analysis routines, instrumentation modular control via RS-422/232, IEEE-488 to either "nuclear instruments modules" (NIM) or "computer automated measurement and control" (CAMAC) crates, integration with the AppleTalk local area network, direct access to the LaserWriter printer, access to AppleShare volumes, etc. Additionally, I hoped that the program structure could be adapted to other laboratory or group needs — an attempt to componentize the software. I also had ideas of laboratory and office computers on the local area network being able to exchange job requests to run experiments or process data remotely. The idea of remote operation over the local area network was shelved as the technology was not there yet. I revisited the remote operation idea later when I worked at Sandia National Laboratories (SNL) and was then able to include it in a working system which we used in various field tests.

I eventually bought myself a Macintosh II computer for my own use. This allowed me to work on computer and software issues which were separate from the laboratory hardware for my own learning, writing (a big improvement over the manual typewriter), and entertainment. Naturally, when the Macintosh II for the laboratory arrived and Ahmed realized it was here, he got all excited and basically asked if I could have everything done by the following Monday. How could I say no?

So I dove into what was essentially a black hole for me. I only knew BASIC, Fortran 77, and RPN at the time, so I had to quickly pick up programming languages from scratch, such as Pascal and C. I had to learn an unfamiliar programming paradigm — GUI and toolboxes. I also had to get up to speed with low level details I did not yet understand, like the Motorola 68K assembly.

I ran to the Caltech bookstore and tried to find some useful Macintosh programming books. Some sample Pascal code for the old SimpleText application was available, so I tried starting with that. Our software development environment was called Lightspeed C, but the Macintosh toolbox documentation was in Pascal, so I had to reverse engineer the Pascal code into equivalent C-code. Adding to the struggle was trying to understand how to integrate low-level assembly code for instrumentation control.

I went into final exam mode for almost a week and came back with version 0.1 of the program. It was a total kludge — I had a baseline GUI set up which allowed commands to be sent to the stepper motors and CAMAC crates, retrieved and displayed data graphically in a window, saved new data to a file, and opened existing data files. However, if you opened a new window, it would wipe out your existing data view since I did not have multiple independent and overlapping windows working yet. There was not much in the way of data analysis routines, except for linear and single exponential decay regression analysis. The program crashed after a while with everyone's favorite system error types 1, 2, or 3 (all memory-related). It would be a little while longer before I got the software stabilized. Then I started adding other features like enhanced GUI controls for data acquisition and analysis, additional analysis algorithms, dynamic selection of data file portions for plotting and analysis, data file transfers across networks, printing, audio, more software componentization, and so on.

I learned a few things after going through the process of writing version 0.1 of the program. One was to check out what comes in the box with your software. As it turned out, Lightspeed C provided sample codes for TextEdit that are already transcribed into C, which helped me find and rectify many of my initial user interface-related mistakes. Another is that software development is hard and a neverending process.

So, I was ultimately able to move the group into the realm of the microcomputer, GUIs and toolkits, laboratory automation, laser printers, and early implementations of local area networks. The laboratory and data acquisition software was flexible enough to be adapted to multiple laboratories. With desktop publishing allowing the creation of figures and printing directly to transparencies on the networked laser printer, we saved quite a bit of money on figure preparation for manuscripts. It would not be long before we moved into desktop multimedia.

It felt like a massive amount of work, but Ahmed appreciated the final results. I remember Ahmed initially being hesitant about getting the LaserWriter printer because the initial expense was US$6000. But a few publications later, he quickly realized it was one of his better investments. You saved money by not having to get expensive glossy prints made for figures, and the printer was on the network so all group members could use the printer for their figures. Another consequence was that Ahmed would subsequently ask me to help prepare figures and presentations whenever special needs came up.

The most prominent of these special needs was the Watson Lecture that Ahmed gave to discuss our current work in femtochemistry. He wanted a short animated film to show the key ideas of the experimental method and our results with ICN (cyanogen iodide) and NaI (sodium iodide). Marcos Dantus, Martin Gruebele (a postdoctoral scholar from University of California, Berkeley), and I met with Ahmed several times for intense discussions about the screenplay, and then we went off to fill in the details. Marcos arranged for Caltech's audiovisual department to take a video of Femtoland 1 in action (their crew learned the hard way that a stray laser shot can burn a camera detector) and made 2D animations of atoms moving. Martin did the numerical calculations of the 3D potential energy surfaces and provided appropriate background music. Using Martin's numbers, I created the visualizations of the atoms moving on the 3D potential energy surfaces and put everything else together in a single presentation. My apartment served as the post-production studio!

Towards the end of my graduate career, laboratory software development tools had matured enough that we started to look into using LabVIEW. LabVIEW tried to make it easier to interface instruments with plugin cards by abstracting away the need to directly deal with assembly. LabVIEW became useable with version 3, and by then I was working at SNL on larger-scale laboratory systems. LabVIEW allowed me to focus on custom software code rather than recreating everything from scratch.

Other Anecdotes

Ahmed organized Linus Pauling's 85th birthday celebration at Caltech on 6 March 1986. That day was declared an academic holiday, so the group went to hear the various technical presentations and later had lunch with Linus at the Athenaeum. It was an absolute privilege to meet Linus and, being in awe, I had no real idea of what to say. But I was able to keep up with him and got his autograph on my copy of his famous book, *The Nature of the Chemical Bond*.

However, I did get my 15 seconds of fame by being in a picture (upper left corner) with Linus in the *LA Times*. In that same article, the reporter noted that I, Marcos, and fellow Zewail graduate student Issac Xavier "vowed to take Vitamin C to match their hero's energy." I do not remember making that vow, but I was likely stuck in a state of mental trance. A copy of the *LA Times* article is shown Photos 9.2 and 9.3.

It is probably fair to say that each group member had some non-scientific traits that are remembered by their contemporaries. I was known for being an avid reader of military and political history (and I still am). I recall once that the group was socializing at the Athenaeum and Ahmed could not help but notice me reading an issue of the U.S. Naval Institute "Proceedings" discussing the U.S. Navy's Maritime Strategy in the 1980s. Talk about a change in topic from femtochemistry!

But probably what Ahmed and most group members of the time also remember me for is my habit of eating a head of lettuce as my main meal (either lunch or dinner) for almost three years, and I only ate twice a day. At that time, you could get a big solid head of iceberg

L.A. Times; March 6, 1986

'I Could Probably Have Talked About Myself More Effectively and Better.'

—Linus Pauling

85 and Going Strong, Pauling Hears the Cheers

LOU MACK / Los Angeles Times

By MARY BARBER, *Times Staff Writer*

When he was 21, Linus Pauling entered Caltech where he would spend 42 birthdays before saying goodby, still stepping off to the music of a different drummer.

When he returned to Pasadena last week, it was to celebrate his 85th birthday and to hear himself hailed as the greatest chemist of the 20th Century, one of the greatest scientists in the world and the only person ever to be awarded two unshared Nobel Prizes, and really deserving a third.

At his 85th birthday celebration last week, Pauling was called "the true father of molecular biology," "the man most responsible for the nuclear test ban treaty," "an enormously effective lecturer with a seraphic smile" and "a slightly enigmatic genius."

Pauling, as noted for his self-confidence and humor as his genius, replied that: "It's been a great experience, hearing all these talks today. I could probably have talked about myself more effectively and better."

Academic Holiday Declared

The seraphic smile never waned, only widened after the daylong tribute that filled Beckman Auditorium and the Athenaeum with scientists, students, family and old friends.

Caltech declared an academic holiday for the event that was staged by its Division of Chemistry and Chemical Engineering and headed by chemistry professor Ahmed Zewail. Most of the speakers were fellow scientists who expanded on Pauling's historic findings in chemical bonding, the structures of molecules, antibodies and the structure of proteins. His book, "The Nature of the Chemical Bond," was the masterwork that won him the Nobel Prize in science in 1954.

Other speakers referred to Pauling's efforts toward world peace and against nuclear arms, which pitted him against anti-Communist forces in Congress but won him the Nobel Peace Prize in 1962. Pauling left Caltech two years later, having been demoted by Caltech trustees who were sensitive to the glaring publicity he attracted. He had been removed as chairman of chemistry and director of two labs.

Please see PAULING, Page 3

CALTECH

Prof. Norman Davidson projects a cartoon of Linus Pauling.

Photo 9.2. *Los Angeles Times* article about Pauling 85th Birthday Celebration at Caltech on 6 March 1986, Part 1.

lettuce for a dollar from the nearby Ralph's grocery store on Lake Avenue. It was one of the ways I made the finances work every month. Although I was able to balance the books for the first few years, for the last few years I was always on a slow bleed. So I would be munching away, whether I was in the laboratory or in a group

L.A. Times; March 6, 1986

During his 85th birthday salute at Caltech, Linus Pauling signs an autograph for Margaret Thurmond of Pacific Palisades.

LOU MACK / Los Angeles Times

Pauling greets his grandson, Linus Kamb, 21, of Pasadena, who is studying architecture at the University of California, Berkeley.

PAULING: 85th Birthday Brings Cheers at Caltech

Continued from Page 1

Speakers made only fleeting references to Pauling's latest, and most controversial, work in which he claims that large doses of vitamin C prevent a variety of illnesses including cancer.

In a brief talk that was riddled with inside scientific jokes far beyond a layman's comprehension, Pauling said, "Of all the places in the world I might have gone to, Caltech gave me the best education, the best preparation for the work I was to do. The Caltech years were the greatest in my life."

Then he bounded down a flight of stairs from the Beckman stage, up the aisle, and into eager arms.

Among those who greeted and paid tribute to him were Nobel winners Francis Crick of the Salk Institute in La Jolla, Henry Taube of Stanford and Willy Fowler, a physicist who in 1983 became Caltech's 20th and most recent Nobel laureate.

Pauling was hugged, patted, followed and adored all day. In a scene reminiscent of 1960s peace marches when he was frequently pictured in eccentric clothes and settings, Pauling happily posed with a pretty girl kissing him.

Recalling their work together, Norman Davidson, Caltech's Norman Chandler professor of chemical biology, said Pauling was "the leading chemist in the 1930s, enough to entitle him to a Nobel prize then, but the committee was timid.

'Don't Say It's Crazy'

Davidson, a Pauling associate for almost 20 years, added, "If Linus Pauling has an idea, it's audacious, but don't say it's crazy. All his ideas are interesting; they don't all turn out to be right."

The tributes came from a variety of sources . . .

- Caltech President Marvin Goldberger called Pauling "vibrant—sometimes overly resonant—in chemistry."
- Ken Trubler, a Pauling student who became professor of chemistry at UCLA, called the guest of honor "one of a handful of men who made Caltech what it is today."
- B. B. Sharma, another former student and professor of chemistry at Pierce College and Cal State Los Angeles, said Pauling's mind "is doing more today than the minds of brilliant men who are many years younger."
- Geoffrey Dolbear of Diamond Bar took a day off from his work as a chemical engineer in Brea and came bearing his old college text from 1961—"The Nature of Chemical Bonds"—for Pauling's autograph. "This is the book that tied it all together," Dolbear said.
- Linus Kamb, 21, an architecture student at the University of California, Berkeley, said, "I knew he was famous and all, but to me he's been just my grandfather and I always loved being with him."

Three Caltech doctoral candidates in chemistry, Larry Peng of St. Louis, Marcos Dantus of Mexico and Isaac Xavier of Brazil, kept pace with Pauling all day and vowed to take vitamin C in order to match their hero's energy.

'Some Kind of Dream'

Xavier said he bought a Portuguese translation of "The Nature of Chemical Bonds" when he was "just a kid in Brazil, just for the excitement of reading it. To me, it's some kind of dream, meeting Linus Pauling."

Several autograph seekers clutched copies of Pauling's latest book, "How to Live Longer and Feel Better," written in his home in Big Sur and published last month.

In a talk titled "Life with Linus," Edward Hughes, Caltech senior research associate emeritus in chemistry, said Pauling predicted at his last Caltech birthday party in 1975 that his next official celebration would be in 1991, when he will be 90.

"That's his first serious error in prediction," Hughes said. "It came five years early. Everyone is invited back on Feb. 28, 1991."

Photo 9.3. *Los Angeles Times* article about Pauling 85th Birthday Celebration at Caltech on 6 March 1986, Part 2.

meeting. Ahmed certainly took notice as one of the first things he told me after graduating was that I "can now afford a better meal." I did not always see Ahmed as the most sentimental type of person, but with that message I knew he cared. A copy of Ahmed's letter to me is shown in Photo 9.4.

CALIFORNIA INSTITUTE OF TECHNOLOGY

Arthur Amos Noyes Laboratory of Chemical Physics, Mail Code 127-72
Pasadena, California 91125

AHMED H. ZEWAIL

LINUS PAULING PROFESSOR
OF CHEMICAL PHYSICS

Telephone: (818) 356-6536
Telex: 675425 CALTECH PSD
FAX: 818-792-8456

December 3, 1990

Dr. Lawrence Peng
IBM Corporation
Department #323, Building #967-2
1000 River Street
Essex Junction, Vermont 05452

Dear Larry:

Thank you for your letter of November 17, 1990. I am glad you are settling in, and you are still eating lettuce for lunch! You can now afford a better meal.

Keep up the good work.

Best wishes,

Ahmed

Ahmed H. Zewail

AHZ:lm

Photo 9.4. Ahmed's letter to me as I began my first job at IBM after leaving Caltech, advising me that I "could afford a better meal."

There were non-scientific group activities every once in awhile. It might be a group lunch, going to the Athenaeum for a beer (albeit non-alcoholic for me), or something else. But two highlights involving Ahmed's non-serious side immediately come to mind.

First was the chemistry and chemical engineering division's Christmas party in 1984. Typically, first-year graduate students have to put on a show. But in addition to us trying to do our best rendition of Ghostbusters, the party organizers also had a belly dancer and got Professor Harry Gray to do his own version of belly dancing. While Harry was dancing, Ahmed ran up to give him a tip!

The second was Ahmed and Dema's wedding reception at the Athenaeum. The entire group was invited to attend. Again, there was a belly dancer present, and Ahmed selected fellow graduate student Earl Potter to perform. Despite other members of the group cheering him on, Earl did his best impression of smiling and largely standing still.

Winding Down, the Real World, and Ahmed's Influence

Upon reflection, I was certainly blessed to have the opportunity of doing my graduate studies at Caltech. While at Caltech, I was able to participate in interesting scientific projects using the cutting edge technology of the day and meet and hopefully learn something from other masters of their fields (e.g., Linus Pauling and Richard Feynman). I was able to set up the laboratory foundations for the Zewail group and move them into the desktop computer age, automate laboratory data acquisition and analysis, and initiate the use of desktop multimedia.

For various reasons, I decided to switch to the corporate or government route after graduation instead of pursuing an academic career. My reasoning at that time may have been idealistic and naive, but I have no regrets. I wanted a life, to possibly start a family, and stability. The prospect of likely two more postdoctoral stints plus another potential five years trying to get tenure was not really appealing. It did not help that the academic market was very tight at that time. I had been going for 90-100 hours a week for most of my graduate career and was starting to feel burned out. But despite the hard work and long hours, it was a dream come true for me to attend Caltech and I still proudly carry that association.

As it turns out, the non-academic job market was fairly tough, so I was still looking for work as the official 1990 graduation ceremonies were approaching. In addition, Ahmed wanted to have complete documentation prepared for 047A as part of my thesis. This meant that although I had passed the final exams, I would not be able to finish all the documentation on time. But fortunately, Ahmed hired me as a consultant in order to finish up the writing. Meanwhile, I succeeded in my job search and I left campus in October 1990. I returned to campus for graduation in June 1991 and subsequently also made short visits if possible when I was in the Los Angeles area. I also did a few journal submission reviews for Ahmed while he was editor of *Chemical Physics Letters*.

I reconnected with Ahmed in January 2000 as Caltech's Division of Chemistry and Chemical Engineering invited group members back to campus for his Nobel Prize celebration. Here is where I found out that Ahmed insisted on inviting all former group members. It was another reminder of how he viewed us as part of an extended family. I have always considered it a privilege to be able to say that I played a small part in the research that would earn Ahmed the 1999 Nobel Prize in Chemistry.

I started my post-Caltech career by going in a different direction and my focus has changed over the years depending on circumstances and funding. Yet, the temperament and many of the skills I honed while working at Caltech have continued to be influential to this day. For me, Ahmed's influence and counsel is less related to the type of work I do, but more towards how I do it and approach it.

At IBM and SNL, I was participating in the implementation of laser-based instrumentation for trace detection in semiconductor manufacturing, waste disposal, and material degradation. For SNL, the laboratory system was packaged into a mobile unit that was taken to several pilot plants. System and laboratory instrumentation software were crucial for those systems and had to be scaled up for use in pilot plants. Here, what I did at Caltech for software and instrumentation literally served as the starting point, and I was able to incorporate features that new technologies finally enabled me to do. The software and system design would be applied to other SNL laboratories for applications in mechanics and reaction kinetics. I would be consulting for these laboratories over the lifetime of these projects.

SNL is also where I further developed the knack for troubleshooting all laboratory and departmental computing issues. Programming gave me an understanding of how the internals of computer hardware and software worked, which made it easier to track down obscure issues. That ability would be crucial as funding and project priorities changed by the late 1990s and ultimately resulted in the start of a career change to institutional and executive-level information technology at Lawrence Livermore National Laboratory. The career change was complete with the move to the Coast Guard and now University of California, Berkeley, where I am currently responsible for information technology for the highest levels of management.

Part of my current career success is certainly attributable to particular traits instilled or amplified by Ahmed. What I saw in Ahmed was intensity, the ability to focus, and a refusal to give up. That he thought things through in the big picture but still worked on getting the details right, had the confidence that problems were ultimately solvable (and that we students could solve them too), and never lost confidence even if we hit an apparent impervious wall. Another exemplary trait which Ahmed did not always articulate explicitly is that one should never stop learning and trying to improve. If you have trouble ascending one mountain, the onus is on you to learn whatever is necessary to get to the summit. Once there, you still see how much more there is to be done. He expected us to work hard, but he would always set an example himself.

Those traits have enabled me to handle multiple crises while maintaining my sanity. The executive level is where crises have wide impact or get blown out of proportion, so containment is often of paramount importance. You must see both the big picture and the small details, navigate political minefields, and convey continual optimism and confidence to gain executive-level support. You have to have a plan B or C ready and be available 24/7. I have colleagues who have said that no detail gets past me, that I do not know how to give up, that I need to slow down, or that even if I'm on call 24/7, I can somehow still find time to help them.

With his untimely passing in August 2016, I had the opportunity to ponder how influential the field of femtochemistry has become since we took those first steps with Ahmed at Caltech. Ahmed always said that we dream big at Caltech, and femtochemistry fits that description. So it was an absolute certainty that I would attend the Ahmed Zewail Memorial Symposium in January 2017 at Caltech. This event was the appropriate avenue for me and perhaps also for other Zewail group members past and present to celebrate Ahmed's legacy and say goodbye.

It was gratifying that a larege nuember of group members were able to attend. It was great to get reeacequainted with people you once worked with, meet your predecessors and successors, and speak with members of Ahmed's immediate family. The group's direction has changed over the years, but each era builds on top of the accomplishments of the previous. Certainly, that was reflected in the quality and chronology of the various presentations.

In closing, I would like to thank Ahmed for allowing me to be part of his group and scientific family. I hope in my own way I have absorbed the best of his enthusiasm for science, knowledge, and life. Together with guidance from my family and other friends, mentors, and teachers, I have been provided with a foundation that guides me through the occasional turbulence of life and career. Even as Ahmed now rests in peace, I am sure he is thrilled that his colleagues and students continue to carry forward his legacy.

Dr. Lawrence W. Peng grew up in St. Louis, Missouri and went to Purdue University for a B.S. in chemistry in 1984. He then went to Caltech in 1984 and graduated in 1991 with a Ph.D. in chemical physics working with Professor Ahmed H. Zewail.

After Caltech, he worked on optical spectroscopic techniques for chemical trace detection while at IBM manufacturing in Burlington, Vermont from 1990 to 1992, and later at Sandia National Laboratories at Livermore, California from 1992 to 1999. While at Sandia National Laboratories, he also specialized in implementing computer control of laboratory instrumentation and scientific computing. He subsequently shifted his focus to scientific and business information technology and moved to Lawrence Livermore National Laboratory in 1999 to support their executive management. Following major changes at Lawrence Livermore National Laboratory in 2008, he then did information technology work for the United States Coast Guard (USCG) at Coast Guard Island in Alameda, California, serving USCG District 11 and USCG Pacific Area. Since 2012, he has been at the University of California, Berkeley and is currently responsible for the information technology of the university's executive management and staff.

10 Under the Influence of Ahmed Zewail

Marcos Dantus*

In my Ph.D. thesis, I wrote: "Zewail knew how to take my enthusiasm for science and amplify it by many orders of magnitude." That sense of excitement for what I was doing and for what I was about to discover kept me awake many nights as a graduate student at Caltech. This is why, when remembering those days, I decided to entitle this account as I did. I first met Zewail in the spring of 1986 when visiting Caltech; I was making sure it was the school where I wanted to do my Ph.D. studies. That first meeting was very memorable because I could immediately see the passion Zewail had for scientific discovery. When he told me that his goal is to observe a chemical reaction in real time (as if it were occurring in slow motion using a fast camera), I told him that I would love to build the laboratory for those measurements. That meeting was decisive — I had to attend Caltech and join Zewail's research group. That fateful decision led to eight years filled with excitement, first as a graduate student and then as a postdoctoral scholar.

The environment in Zewail's research group was very special. Everyone was so knowledgeable and competitive. When we were given a project, we would focus all our energies towards getting results. The majority of the time, we learned by interacting with other members of the research group. I remember when I joined the group,

* Graduate Research Assistant/Postdoc.
Email: dantus@chemistry.msu.edu

Zewail told me to learn as much as possible from Peter Felker. When I asked Peter to show me the laboratory, he walked me to the laboratory — the lights were off as they should in a functioning spectroscopy laboratory — and told me, "We have an Argon ion sync-pumped mode-locked cavity-dumped dye laser intersecting at right angles with a supersonic molecular beam. We collect molecular fluorescence at right angles through a spectrometer and detect the signal using time-correlated single photon counting. Do you have any questions?" Frankly, I did have lots of questions. I had never seen a system like the one he was using. However, I realized it was my responsibility to read the papers and figure it out.

I learned how to run the lasers in that laboratory by watching Spencer Baskin set up fluorescence lifetime scans, where some of the very first rotational recurrences were observed. Zewail asked me to work on determining the number of solvent molecules required to enable an excited state charge transfer in dimethylamino-benzonitrile. That project required, in addition to the other systems, a tunable dye laser pumped by an old nitrogen laser. I happened to have used one of those as an undergrad. I finished that project about a month after arriving at Caltech. Zewail could not believe the speed with which I completed the project, so he asked Lawrence Peng to repeat all the measurements. After that, Zewail talked to Professor Kenneth Eisenthal and decided to write a joint paper which included data from their laboratory as well. In the meantime, I was given a second project; this time I would work with Spencer Baskin. The second project was elegant because it explored how the length of an alkane chain in the para position of aniline affects intramolecular vibrational energy redistribution (IVR). This project was interesting because it was difficult to predict from the fluorescence spectrum how the fluorescence lifetime would look. Many of the molecules showed restricted IVR with very nice quantum beats; however, others just showed some bi-exponential decay. As this project dragged on for a few weeks, I remember telling Spencer that I would do everything in my power to get it done. I didn't sleep for two days and completed the data acquisition. While we were writing that manuscript, Zewail called me and asked me to design a femtosecond laser laboratory.

Photo 10.1. This picture was taken in the summer of 1986 as Zewail and I were contemplating the conversion of an old x-ray facility into a femtosecond laser chemistry facility. The laboratory was ready by the end of November 1986 and we had results (the ICN (cyanogen idodine) experiment) by February 1987.

That was the moment I had been waiting for! I include here a picture (see Photo 10.1) of Zewail and I contemplating a laboratory full of old X-ray diffractometers used by Professor Linus Pauling. You can see us imagining how we will transform it into a state-of-the-art femtosecond laboratory to usher in a new era in the study of chemical bonding, going from static structures to unraveling the dynamics of chemical bond cleavage and formation in real time.

The design and construction of the laboratory took place in the ultrafast timescale, Zewail's favorite, and within months we were measuring molecular dynamics that had never been accessible to mankind. In that sense, the early days of femtochemistry were magical. Every compound we placed in front of the laser led to a breakthrough scientific discovery. I am incredibly thankful for the trust Zewail placed in Mark Rosker and me on the design and construction of a million-dollar laboratory. The freedom he gave us in the laboratory allowed us to learn by doing. His enthusiasm for new results was sufficient motivation to run the laser day and night. Once all the data had been acquired and printed, we would spread it out on Zewail's coffee table in his office and stitch together the scientific findings. I found it fascinating how Zewail would place our findings within a greater scientific picture that started with what was known and then move on to what had been found. Finally, he would consider the far-reaching implications for our findings.

When we were working on the first two femtochemistry papers, known in short as FTS I and FTS II, we would start in Zewail's office sometime around 5.00 p.m. We would break at some point to grab some food and reconvene at 10.00 p.m. in order to keep working. We would often carry on working until 2.00 a.m. Walking on the streets at those times was not the safest; Mark Rosker once narrowly escaped a gang while going to enter the latest changes to the manuscript and printing a fresh copy. Even with the lasers working perfectly, data acquisition for that project required more than 24 hours, so this required us to make sure the laser did not lose mode-locking. In those days, the dye-based colliding pulse mode-locked laser would stop working every so often. Although we had programmed the computer to stop taking data when the laser stopped, we wanted to get the laser back in action as soon as possible. In those long days, Mark and I found ways to have fun in the laboratory while keeping an eye on the laser.

Scientific discussions with Zewail were always wonderful. In addition to talking about ongoing research, we would discuss the different seminars and colloquia at Caltech. I enjoyed hearing new perspectives and learning about new powerful research tools being developed and would share those with Zewail. The discussions were always best when they involved coffee or food. If the conversations took place in his office, then I would join him in enjoying coffee with cinnamon. If we were out for lunch, then it was usually Italian food.

In a typical scientific conversation, Zewail would focus intently on a specific aspect of the scientific process being discussed. It was clear he wanted to understand every nuance, and for this he had a unique approach. He would create a model in his mind that contained the most essential aspects and ignored every bit of additional information that was not critical. For example, when thinking of a chemical reaction, he would focus on the primary steps and ignore all the complexities that would confuse the fundamental process he wanted to understand. Note that while this approach appears naïve to an untrained observer, Zewail was able to simplify complex problems, get to the heart of them, and once understood he would then bring the complexities back in. That approach was followed in almost every research project tackled by the group. There were many beautiful progressions such as the study of IVR, especially the article that explored the length of an alkane chain in para-substituted aniline molecules; solvation with an increasing number of water or methanol molecules and its effect on charge transfer in dimethylaminobenzonitrile; increasing complexity in bond cleavage from ICN to predissociation in NaI followed by dissociation from a saddle point in HgI_2. By the time femtochemistry was turning 10 years old, Zewail was addressing common organic chemistry mechanisms and also making good progress on ultrafast electron diffraction where motions on the entire molecular structure could be tracked in real time.

It is worth noting that the atmosphere in Zewail's group was very intense. Within hours of starting a project, Zewail wanted to hear about results. As soon as you had results, he wanted to see the manuscript. Every project seemed to be the most important mission in the world and there was nothing more important than completing the work. In those days, you would find graduate students sleeping on the famous sofa in one of the offices. There was very little communication between students working in different laboratories. I remember a time when I cut myself in the laboratory. I wrapped my finger in paper towels and went to Zewail's office to explain that I may be absent from the laboratory for a couple of hours while I go to the hospital. Concerned, he looked at the dripping blood and said, "Marcos, let me take you to the hospital right away." It was reassuring that, although he valued and prioritized intense work, he was also concerned with my wellbeing. But after getting treatment from the hospital, it was straight back to the laboratory.

The atmosphere started changing when postdoctoral scholar Robert (Bob) Bowman joined our research group. Bob had the most remarkable way of making friends. He would not accept that people were angry and remained upset for a long time. I remember a day when Bob had a discussion with Mark Rosker about a half-wavelength plate we had been using in the laboratory. Mark left the laboratory very angry and slammed the door. After a couple of hours, Bob suggested we invite Mark for lunch. I was surprised. Bob asked Mark to join us and Mark said he was not interested. Bob smiled and poked Mark on the side and made him laugh, and kept asking, "C'mon Mark join us!" Mark finally relented and joined us. Bob kept working his magic with the rest of the group. When Zewail was awarded the King Faisal Award, Bob suggested that we have the King Faisal Invitational Golf Tournament amongst all the group members. I imagine it was the first time the entire research group went out during a weekday to have fun.

Bob and I had the best time working together in the laboratory. We worked together from 1989 to 1990 and published eight projects on HgI_2 and on I_2. We would have easily finished multiple other projects but unfortunately Zewail did not allow us to work together any more. The reason must have been that Zewail wanted us to work with other group members. I then had the privilege of working with postdoctoral researcher Maurice Janssen. Working with Maurice was very important for me. I remember at some point, while working on our first project on the new molecular beam, Maurice decided to stop the project to do maintenance. Used to the high pressure of the group for so long, the concept of stopping and taking care of the equipment was completely foreign to me. Maurice explained that if we took care of the equipment, we would get very good data and if we didn't, we ran the risk of ruining the vacuum system and that would result in weeks of down time. This was an unexpectedly important lesson for me.

As the popularity of femtochemistry grew, Zewail's group became a major destination for postdoctoral scholars from many parts of the world. I had the privilege of working with Dr. Gareth Roberts on a perspective paper on femtochemistry followed by a paper with Dr. Martin Gruebele on inverting time-resolved data to obtain a potential energy surface. Both of them taught me to appreciate the deeper meaning of the experimental data we were collecting. With Martin, I got to work on a computer animation that was the highlight of a public address by Zewail on Femtochemistry. That address, given at Caltech, was an opportunity to reflect on what was being accomplished and its implications for chemistry and other sciences. We used beta-versions of computer software that would later evolve into the platforms for Adobe and multiple other companies for such projects. In those days, it would take days to render just one minute of video. Zewail liked the animation very much and we called it *The Birth of Molecules*.

During the summer of 1990, Professor Richard B. Bernstein, Zewail's friend, colleague, and collaborator, had a heart attack while in Moscow and passed away soon after at a hospital in Helsinki. Bernstein had a tremendous influence in Zewail's research. In the early days, it was Bernstein who suggested to Zewail that he should focus his knowledge of picosecond laser sources on molecular dynamics rather than on energy transfer. Bernstein insisted that shorter timescales would be needed to understand chemical bonding. Their interactions grew stronger at a conference that Zewail organized at Caltech in 1980. An article from that time by Zewail mentions how, with fast enough pulses, one may be able to beat IVR. Bernstein must have influenced the first femtosecond research projects. Bernstein had studied ICN, NaI, methyl iodide, and HgI_2 in his laboratory, and he was also keen on the problem of inverting time-resolved spectroscopic data to obtain the potential energy surface behind the observed dynamics. Zewail and Bernstein jointly wrote a remarkable and influential review and they may have been destined to share a Nobel Prize. Alas, Bernstein's heart attack took everyone by surprise. Zewail endured a very difficult time for the rest of 1990.

Upon my graduation in 1991, it was time for me to consider a postdoctoral position. At that time, Zewail was considering the development of femtosecond time-resolved electron diffraction. As part of my candidacy exam and Ph.D. defense, I had proposed doing electron diffraction in molecular beams and it seemed to me like a perfect opportunity to stay in Zewail's group. For this project, Zewail allowed me to work quite independently; he gave me a budget and two outstanding colleagues, Scott Kim, a postdoctoral researcher with a Ph.D. in electrical engineering, and Chuck Williamson, a new graduate student. The three of us set out to design and build the first femtosecond electron diffraction system for studying chemical structures with ultrafast time resolution. Scott kept us from getting hurt by the 15 kV acceleration voltages and Chuck kept making improvements on the data acquisition software, sometimes only hours after I suggested them. This work established many of the details required to accomplish ultrafast electron diffraction.

My hard work in the research group did not go unnoticed; Zewail successfully nominated me for the Clauser Prize at Caltech, which is given to the Ph.D. thesis considered to have the potential of opening new avenues of human thought and endeavor. He also nominated me, successfully, for the Nobel Laureate Signature Award given by the American Chemical Society for the most significant Ph.D. thesis in chemistry. In the picture (see Photo 10.2), Zewail and I are happy for the line of successful projects completed between 1985 and 1991. The hard work and ingenuity had paid off handsomely.

Photo 10.2. This picture was taken at the award ceremony during the American Chemical Society meeting held in April 1992.

After leaving Caltech, it was time for me to grow as an independent professor. Zewail was quite supportive. Insofar as he had sufficient time, he always wrote me letters of recommendation for different awards and for tenure when I requested them. Once I left Caltech, the femtochemistry conference was the place to meet with Zewail and many alumni from the group. Zewail wanted to make sure the conference allowed ample time for scientific discussion and he also wanted it to be different from other conferences. Instead of multiple parallel sessions with hundreds of 15-minute talks, Zewail wanted a conference with no parallel sessions where thoughtful presentations were made and discussed. At these conferences, Zewail's talk was always a tour de force. While most people completed a few projects since the last conference, Zewail typically covered ten or more projects and showed incredible new results.

The Nobel Prize for Zewail was only a matter of when. When it was announced in 1999, I was driving to my office and got so excited. I called Zewail but he was in the shower, so I had to call again 10 minutes later. Frankly, when we were working on those early projects, we had a sense that we were doing very high impact science that would likely be recognized with a Nobel Prize. The Nobel Prize had a profound impact on Zewail. Most notably, he became even more energized and focused on science. Unlike many Nobel Prize winners, Zewail immediately set to work on his second Nobel Prize. He wanted to bring ultrafast time resolution to electron microscopy. He wanted to see biological machinery at work with nanometer spatial resolution. In parallel, Zewail felt that he had an enormous responsibility to society. Zewail started giving lectures related to science education and understanding between the Western World and the Middle East. He became an avid proponent for scientific education, especially in countries with fewer resources, and he also became a crusader for fundamental scientific research. These efforts were noticed worldwide and he became appointed by President Obama as a scientific advisor and science ambassador. It is unbelievable that the productivity of the research group never suffered while he kept adding responsibilities to his workday.

At the femtochemistry conferences, there was always an opportunity to invite Zewail for a meal or coffee and to talk about one's career and science. My favorite such conversation took place in Paris when Zewail invited me for dinner at Le Fouquet's. The amazing meal was followed by coffee along the Avenue des Champs-Élysées. It was on that occasion that Zewail told me that he was very glad of all my successes on the development of pulse shaping and laser control of nonlinear optical processes. He was glad that we had figured out the science behind coherent control of chemical reactions, especially because during that time there had been many reports that lacked a satisfying scientific explanation.

My conversations with Zewail continued with every femtochemistry conference until he missed one due to medical reasons. I remember talking to him and telling him that just like he had shown everyone that no project was insurmountable, he would likewise overcome his illness. We met again in March 2015 while I was visiting Caltech for the National Academy of Inventors meeting (see Photo 10.3). It was then that Zewail told me that I should plan the

Photo 10.3. This picture was taken in Zewail's office and includes my wife, Debora Dantus. The meeting took place in March 2015. It was at that time that we started planning for the 2017 Femtochemistry Conference in Mexico.

2017 Femtochemistry Conference in Mexico. At that time, Zewail was in great spirits as can be seen in the picture. My wife and I spent some time with Zewail while planning for the conference in 2017 and catching up about our families. It was a very sad moment when I learned of his untimely passing in August 2016. This is a tremendous loss for science and scientific education as we have lost an important role model. I hope these memories, and the many more in this book, illustrate the multidimensional impact that Zewail had in this world. Zewail is a great example of how life is not about what we take from it but what we leave behind.

Marcos Dantus received B.A. and M.A. degrees in chemistry from Brandeis University. At Caltech under Ahmed Zewail, Dantus received his Ph.D. (1991) and also served as a postdoctoral researcher (1991 to 1993) where he helped to develop femtochemistry and ultrafast electron diffraction. These developments were recognized by the Nobel Prize in Chemistry in 1999. Dantus has been a professor at Michigan State University since 1993 and is presently a University Distinguished Professor and the Michigan State University Foundation Chair. He has pioneered the use of shaped ultrafast pulses to probe and control chemical reactions as well as for practical applications such as biomedical imaging, proteomics, and standoff detection of explosives. Dantus founded Biophotonic Solutions Inc. in 2004 to develop and commercialize an instrument capable of automated laser pulse compression. The multiphoton intrapulse interference phase scan shapers are now enabling cutting edge research around the world. Dantus regularly collaborates with different branches of the Department of Defense and was invited to the Defense Advanced Research Projects Agency's Scientist Helping America, Arbitrary Waveform Generation, and Program for Ultrafast Laser Science workshops. Dantus has over 225 publications and is a Fellow of The National Academy of Inventors, The American Physical Society, and The Optical Society of America.

11 The Days of Femtochemistry at Caltech

Martin Gruebele*

The late 80s and early 90s were an exciting time in Ahmed's group. Dick Bernstein, who had moved from Columbia to the University of California, Los Angeles, was a frequent visitor in Ahmed's office, and the two had hatched a plan to look at molecular vibrations and transition states in real time. The new femtosecond pump-probe experiment, based on a colliding pulse mode-locked laser design from Chuck Shank's group, had the time resolution to do it. Two floors up was the office of Rudy Marcus, who would soon win a Nobel Prize for his electron transfer theory, but from the time-resolved kinetics perspective, Rudy's refinements of Rice-Ramsperger-Kassel-Marcus theory, which describes how molecules find the "bottleneck" at the transition state, was what we Zewailites were talking about in the basement where the lasers were sitting.

I had just gotten my Ph.D. in November 1988 studying high-resolution spectroscopy of transient molecular ions and clusters with Rich Saykally at Berkeley, after which I moved "down south" to Pasadena a month or so later. In those days, Mount Wilson was still rarely visible behind the Pasadena skyline due to the famous (but now just a memory) Los Angeles smog. I had discussed postdoctoral positions with Dick Zare and Steve Chu in my last months at Berkeley,

* Postdoctoral Research Fellow.
Email: mgruebel@illinois.edu

but my epiphany came while reading theoretical papers by Rick Heller (then at University of Washington) and experimental papers by Ahmed as I was writing up my thesis.

Quantum mechanics fascinated me as a "theory of everything molecular" because it came in two equivalent versions (from the point of view of results) that were, however, philosophically very different: time-independent, where stationary states completely described the system, versus time-dependent, where Heller's wave packets provided a description more reminiscent of classical dynamics, but with added interference and fuzziness. In the 80s, frequency-resolved spectroscopy and crossed molecular beams reigned — Yuan Lee had just won the Nobel Prize at Berkeley for experimental reaction dynamics by scattering. So Ahmed and Dick broke new ground as the experimental counterparts of Rick Heller.

In sub-basement laboratory 047, Marcos Dantus and Bob Bowman were toiling away measuring quantum beats from molecular iodine vibrational motions to the dissociation of HgJ_2. They were continuing the work that Marcos and Mark Rosker had started. I jumped in as the resident theorist for that laboratory and we wrote up a number of papers together, explaining how the femtosecond data could be turned into potential energy surfaces directly, as well as modeling how HgI_2 broke apart and had persistent coherence. Dick Bernstein and I had engaged in friendly theory-theory competition, where he jokingly bemoaned that I was coming up with quantum mechanical solutions before he had finished his classical modeling, and that it ought to be the other way around.

In the other laboratory 048, I joined graduate student Earl Potter, who had worked previously with Norbert Scherer, now at the University of Chicago, and Lutfur Khundkar. We experimented on picosecond spectroscopy of ketene to test Marcus' variational Rice-Ramsperger-Kassel-Marcus theory. A nice paper came out in *Chemical Physics Letters* after a few months' work, although the reviewer (I think Bradley Moore) fortunately corrected a mix-up in the analysis: the observed relaxation rate of a reaction is the sum of forwards and backwards rates, and equal to the forwards rate only when it greatly exceeds the backwards rate. Soon, we were joined by Ian

Sims from England, with whom I ended up having endless religious arguments, I being the resident non-believer. We worked on the reaction of Br+I_2 initiated by the photodissociation of HBr. We got nice data on it from our Antares actively mode-locked amplified laser system, although we were trying to explain it in terms of a ground state reaction surface, when Doug McDonald (now my emeritus colleague at Illinois) later showed that it was really an excited state reaction. In 1991, I also played around with chirping our amplified femtosecond pulse by moving the compressor prisms around. Ian and I looked at I_2 quantum beats in the molecular beam, showing that we could shift the pattern of beats around by chirping the pump pulse. This was a form of early coherent control and I remember discussing this at length with Kent Wilson at University of California, San Diego, who was working on picosecond X-ray imaging and was like a second mentor to me on visits "down south." We never published our data, but Kent made headway with coherent control and several of his postdoctoral researchers went on to work in that field.

Earl Potter was a really nice guy and extremely handy with electronics and assembling control boxes or machining mounts. We had to do a lot of this stuff ourselves because the machine shop and electronics shops at Caltech, while staffed by experts, were totally overworked. However, there was one thing, as I found out, you did not do with Earl: joke around when the Packers lost a game. I found this out one morning when I quipped about the blinking lights on a box he had built and his less than positive reaction took me a bit aback. Someone informed me that the Packers had lost a game. After it was explained to me what "Packers" were, I knew the deal. However, no amount of effort by Earl ever got me interested in football, basketball, baseball, or any of those other team sports.

There was plenty of camaraderie among the various subgroups, which also included Jack Breen, Dean Willberg, and later Jennifer Herek, Luis Bañares, Mike Gutmann, Chuck Williamson, Gareth Roberts, and others. I managed to impress Earl Potter, a pretty big guy, by out-eating him in all-you-can-eat calzones contests. The skinny guys are often the most dangerous, as Gareth Roberts, even trimmer than I was, also demonstrated. We did damage at Edokko all-you-can-eat sushi, Sizzler, and many other places. Ian and I went out on a weekly restaurant tour, trying exotic places in the greater Los Angeles area, such as Hungarian restaurants where waiters would pour digestifs straight into your mouth, or plantain and tongue at a Cuban place. At Gutmann's place, I smoked (part of) my only cigar ever, which kept me nauseous and awake for a whole night. I once took Jennifer Herek to see Disney's *Jungle Book*, as we were both big animated film fans. I was never prouder of my cool than when my car's timing belt gave out en route and I just let it glide to a stop, had someone in a pick-up truck push the car into a parking spot, and then went to the movie without losing a beat. A lot of other beach excursions, mountain hiking trips, and ski trips happened during those years, often accompanied by Ahmed, who was in good shape. Probably the most famous moment was when Michael Gutmann introduced Ahmed to the "Feuerzangenbowle," a rather high percentage German winter beverage we consumed at Mammoth Lakes on a group ski trip.

There were also sad times. One day, Dick Bernstein got very excited about a trip to Russia now that it was opening up with perestroika. He had wanted to go in the 60s but was warned not to do so by the state department. Alas, it was to be his last trip: he suffered a heart attack in Russia, which probably would have been dealt with through routine surgery treatment in the US. But there, it became fatal as he was emergency evacuated to Finland. We were following these events as best as we could in the pre-cell phone era. I nearly broke down when I saw his widow at the wake. Along with everyone else, I was very heartbroken when we heard the bad news, but no one more than Ahmed, who had enjoyed a rare and precious friendship with Dick. I still think about it each time I see Dick's book on molecular beams on my shelf when I turn around in my chair at Illinois.

Ahmed was quite demanding when it came down to work in the laboratory. We were all well aware of this and Bob Bowman, Ian Sims, Marcus Dantus, and the others would time our group dinners and movie events at the local theater on Colorado Avenue accordingly. I usually worked from around 11.00 a.m. to 2.00 a.m., but I did slip once. My labmate was out of town and I was "warming up" the laser (a morning job), so I decided to do the *NY Times* crossword puzzle. As I was sitting on the group couch in the student office staring up close at the puzzle, a finger gently pushed down the newspaper I was holding up, and Ahmed's face emerged from behind. "Martin!" was all he exclaimed with a shake of his head as he turned around to leave the office. I jumped up and assured him this was a singular occurrence, and headed back to the laboratory to tend the laser. Never did things get busier than whenever Ahmed announced, "I am going to a meeting in two days and need the newest data," and we all scrambled to get in some more scans.

Occasionally, we would have parties at Ahmed's house in San Marino and this is where I got to know his wife Dema, and eventually the little boys (now all grown up) that came along. Dema and Ahmed had met at the King Faisal Award ceremony and it must have been love at first sight, for he was married soon. The trip to San Marino was always fun for the group, with interesting Middle Eastern-themed food that was a little more authentic than "Burger Continental," a restaurant where Ahmed and I would hang out a lot because it was just a couple of blocks from the laboratory. Dema was a kind soul, and the postdoctoral scholars and students of Ahmed's laboratory were like a second family for her, and she was the lab-mother. I met up with Dema on many occasions later on, from walking around at the Huntington Library museum, to Nobel celebration dinners, to most recently at a celebration of Ahmed's research in 2017 at the American Chemical Society meeting. I swear, she had hardly changed at all in 25 years.

One major outreach project that Marcos and I did, in an era where outreach was still optional, was to develop a short documentary about femtochemistry. Ahmed specified what he wanted in it and Marcos did his best with the primitive Macintosh animation software of the day to turn it into visuals. I wrote sound effects and music for it as I had purchased a pretty sophisticated MIDI sequencer and synthesizer setup for my apartment, mainly to play Bach organ music and to have orchestral accompaniment playing along with me. As we approached the deadline for the film, the SMPTE synching would not work, and Marcos and I sweated bullets trying to synchronize the film and soundtrack without the image and sound effects wandering out of phase. We finally got it done at the last minute with help from the audio-visual tech people at Caltech, and Ahmed was able to deliver his public evening lecture with the movie. Alas, I missed the show because I was traveling that weekend.

Not long before coming to Caltech, I had discovered the wonder of Mountain Dew, a heavily caffeinated beverage that got me through many nights tending the laser system. I first tasted it, of all places, at the home of the Commanding Officer of Wright-Patterson air force base in Ohio. Said officer was the father of Geoff Blake, whom I had gotten to know as a postdoctoral researcher in Rich Saykally's group, and who was now also at Caltech as an assistant professor. In Ahmed's student office in the basement of Noyes Laboratory, I had piled up a hexagonal close-packed wall of Mountain Dew cans person-high next to my desk. Visitors would invariably comment on

Photo 11.1. Martin and Ahmed in his office *ca.* 1991 during one of our discussions.

it, but Ahmed never raised an eyebrow. He fully understood that adequate nutrition was necessary for night work! Fortunately, I had a good 15–20 minute walk to and from work every day to get some exercise and generally walked whenever I could, saving the tanking-up cost for my 1980 Toyota Tercel (the one with the timing belt problem).

Discussions in Ahmed's office were a particular treat. About once a month or so, he wanted a more thorough progress report and to talk science in detail, and I would come up to the ground floor of Noyes Laboratory (coincidentally, that's also one of the main chemistry buildings at Illinois, my current institution) and see the sun. We usually sat down on a couch in his office (across from his desk outside the frame of the photo) and had lengthy discussions with a lot of gesticulating of my hands, drawing sketches of experiments or data on scraps of paper. Sometimes, we would extend these to lunches at the Athenaeum, where Ahmed wryly commented on the amount of "ze meats" I would consume. The verdict is still out on the funnier accent, when I say "udder" instead of "other" or when he says "meats" instead of "meat." At the Athenaeum, I also got to meet some of the other luminaries on campus. The list includes the indomitable Harry Gray, whom I loved like one loves a crazy uncle who occasionally comes to visit with surprises, and Linus Pauling, who was lecturing about quasi-crystals and happened to be my academic grand-grand-grand-father.

One idiosyncrasy of Ahmed's, when we were planning figures for publications, was that he wanted all the time axes on plots to be in femtoseconds, even if the experiment took a millisecond! I sheepishly confess that I thwarted his wishes once by presenting him with two plots of the same data, one from 0 to 10 picoseconds and the other from 0 to 10,000 femtoseconds. On the femtosecond plot, I had subtly messed up the straightness of the axes and labels just enough that it looked a little disconcerting, but subtle enough so one could not quite tell. To my relief, Ahmed picked the picosecond plot, which is published in the paper we sent off. This is the only time I ever got the better of Ahmed in an argument on figure design. His taste in layout and clarity was generally impeccable and he had a lot of influence in shaping key figures in our publications.

I think Ahmed was a bit disappointed that I did not take over at the European branch of femtochemistry because he had lobbied for me at the University of Frankfurt, and they offered me a C4 position at the ripe old age of 28. Several other postdoctoral scholars and students (such as Jennifer Herek or Maurice Janssen) took care of that by heading back to Europe, so he soon forgave me for staying in the US. I saw him on many occasions, from meetings in Berlin shortly after I left Caltech, to the famous Nobel meeting in 1996 organized by Villy Sundström, Ahmed's long-time co-editor at *Chemical Physics Letters*, and several times at Caltech, during an occasional lecture visit in the 1990s or 2000s. I was very surprised to hear from Dongping Zhong, who joined the group at Caltech after I left, that Ahmed was ill. He recovered for several years but finally lost the battle all too young at age 70, not unlike Kent Wilson, who had passed away even younger more than a decade earlier.

Of course, femtochemistry has made a huge impression in physical chemistry, but most importantly, outside of physical chemistry, in organic chemistry and materials science as well. It really marked a shift in the way chemists think about quantum processes in terms of time-dependent wave packets moving around, instead of stationary states. To be sure, the underlying principles had been well-known since the late 1920s, but there is a big difference between writing down an abstract formalism versus really living and breathing it. It would be rather amusing if a future generation of chemists rediscovers scattering matrices and eigenstate product distributions as their main way of thinking about dynamics, although all these different facets of chemical dynamics are happily married and doing their job these days. We have Ahmed to thank for it in good measure!

Martin Gruebele was born in Stuttgart, Germany in 1964. He obtained his B.S. in 1984 and his Ph.D. in 1988 at University of California, Berkeley. He went on to do femtochemistry in the laboratory of Ahmed Zewail at Caltech before moving to the University of Illinois in 1992. He is currently the James R. Eiszner Professor of Chemistry, Professor of Physics, and Professor of Biophysics and Computational Biology. He is a Fellow of the American Physical, Chemical and Biophysical Societies, as well as a recipient of the Sacker International Prize in Biophysics, the American Chemical Society Nakanishi Prize, and the Wilhelm Bessel Prize, among others. He is a member of the German National Academy of Sciences, the American Academy of Arts and Sciences, and the National Academy of Sciences (USA). He has served as Senior Editor at the *Journal of Physical Chemistry* and as Associate Editor of the *Journal of the American Chemical Society*. His research includes protein and RNA folding, fast dynamics in live cells, vibrational energy flow in molecules, quantum computing and quantum control, optically assisted STM, glass dynamics, and vertebrate swimming behavior. The work is published in over 250 papers and reviews. Martin Gruebele is married to Nancy Makri, with two children, Alexander and Valerie.

12 Remembering Ahmed Zewail

Lynne Martinez*

I was Professor Ahmed Zewail's administrative assistant from 1990 to 1995. I will always be grateful and proud to have worked for him during an exciting time in his career when his research was flourishing. I remember when he first coined the term "femtosecond spectroscopy!" Professor Zewail was always very charming to everyone he interacted with, from his colleagues to staff members. He also had a great sense of humour. I had never worked in a university environment before, so I learned many new things from Professor Zewail — from interacting with other scientists and his world-class research group to university policies and procedures. His former graduate students and postdoctoral researchers have gone on to make names for themselves in academia and in the pharmaceutical industry (Luis Banares, Marcos Dantus, Michael Gutmann, Ahmed Heikal, Jennifer Herek, Maurice Janssen, Rajiv Shah, Dean Willberg, and Chuck Williamson come to mind, to name just a few). During the time I was at Caltech, the Zewail group would always include my daughter Lauren and me at their barbecues at Lacy Park in San Marino or at the Caltech recreation center. It was an opportunity to chat, play a game of frisbee, and unwind. I also had the opportunity to meet Professor Linus Pauling on the occasion of his 90th birthday celebration at Caltech on February 28, 1991.

* Email: lynnem@caltech.edu

In the early 90s, noisy typewriters were quickly being replaced by the whirl of computer terminals at Caltech. I remember doing all of Professor Zewail's letters, class notes, examination papers, and manuscripts on an IBM Selectric typewriter. A manuscript went through many drafts: cutting, pasting, and then typing it all over again. If you made a typo, you had to erase through many carbons. It was distressing, so your typing skills had to be up to par. Now, we can't even imagine life without the computer.

My time at Caltech has been very rewarding, in part due to the professors I have had the honor of working with, and I feel very privileged to have worked for Professor Zewail, a Nobel laureate, and to have played a very small role in assisting him.

Photo 12.1. Crellin and Linus Pauling, Ahmed Zewail, and Lynne Martinez on the occasion of Linus Pauling's 90th birthday celebration at Caltech on 28 February 1991.

13 Challenging Ahmed's Sense of Humor

Michael Gutmann*

Preliminary Remarks

I joined Ahmed's research group in June 1990 as a postdoctoral researcher and spent an exciting time at Caltech until June 1992. Right from the start, I felt that I was given the opportunity of being part of a remarkable group. Ahmed had the ability to bring together people from many different nations who share the same dedication of doing first-rate research. At the same time, this particular group developed a wonderful team spirit leading to friendships that have lasted until today.

It was a real pleasure to work together with Dean Willberg who was a graduate student at that time. We studied picosecond vibrational predissociation of iodine rare gas van der Waals clusters, where we were able to obtain a lot of exciting results. Sharing these results with Ahmed and discussing their implications for theory and reaction mechanism were very special situations where we got carried away by his unique enthusiasm for science.

The Joke

As Ahmed very much enjoyed having a group of people coming from lots of different countries, he developed a tradition that each new

* Postdoctoral Research Fellow.
Email: m.gutmann@liop-tec.com

member of his group had to tell a joke from his home country. After that, all the others were asked to join in. It was my turn, if I remember correctly, when the whole group made a trip to the San Gabriel Mountains. After the barbecue, I was asked to tell my joke. In 1990, political correctness in Europe had not yet advanced to the level it had already reached in the US, so I told a joke that challenged Ahmed's sense of humor a little. The joke was a slightly risqué one about a man and a woman meeting in a German train and having a conversation during the train ride.

From a slight freeze in his face I noted that Ahmed did not like my joke too much. With his unique sense of humor, he gave me a smile and continued without making any further comments.

However, Ahmed remembered very well when some months later it was somebody else's turn to tell a joke. He turned round to me and said, "No, we don't want to hear Michael's jokes anymore." And from that point, the tradition of telling a joke was stopped — at least as long as I remained in the group.

The German Winter Drink

In Ahmed's group each year we enjoyed a very special event. Caltech sponsored a yearly group trip where family members were also allowed to join. The purpose of these trips was not only to enjoy the different environment as a group but also to get together and discuss new horizons of science, proposing exciting science projects, and sharing ideas about the visions that could be new milestones for Ahmed's research group. These sessions were called "science talk" in our group language.

While I was at Caltech, we made two skiing trips, the first one to Mammoth Lakes and the second one to Lake Tahoe. In preparation for the Lake Tahoe trip we discussed various ideas to make it very special. So I proposed to prepare a German winter drink as a starter for one of our science talks. The drink basically consists of mulled wine and rum.

A standard recipe for eight servings is based on the following ingredients: one orange, one lemon, two cinnamon sticks, star anise, two cloves, two liters of dry red wine, 375 mL of rum (100 proof or higher), and a sugar loaf. Peel the orange and lemon, pour their juice into a pot, add the orange and the lemon peel, add the other spices, and then pour the red wine. Heat the pot and avoid boiling. Then place fire tongs or another appropriate metal rack above the pot and put the sugar loaf on the fire tongs. Put some drops of rum on the sugar loaf and light it with a match. Take care that the flame does not go out by adding more rum using a scoop (don't do it from the bottle!). The idea is that the whole sugar loaf melts down into the pot. When this is done, you can start drinking (it is a hot drink and depending on the amount of rum you use, it may contain a good deal of alcohol).

The problem was getting the sugar loaf. The standard loaf needs to be cone-shaped. In Germany, it is available as a single piece, but in Pasadena, we were not able to get a cone-shaped sugar loaf. So Bob Bowman, being a creative experimentalist, got several packs of sugar cubes which he put on a tray and added water until the cubes fused together, forming a rather big lump. He put a copper wire through the lump so that it could later be put on the pot. Then the lump got dry and was carefully wrapped in plastic wrap.

At Lake Tahoe, the first science talk took place in the early evening. So I proposed to Ahmed to prepare the German winter drink for this special evening. Everything worked out fine except that the lump Bob had prepared was enormously large, so it took us a lot of time and a lot of rum to make it melt down into the wine. Ahmed got a little impatient from all the effort we devoted to the drink rather than to the original purpose of this evening. However, he was gentle enough to let it all happen and even tasted some of the drink.

To summarize, the first science talk did not exactly run the way it was intended to. The discussions went rather short and the contents were not as profound as they should have been. I am not quite sure whether this was caused by the delay in starting the serious discussions or the copious amount of rum that was needed to melt the sugar lump. Maybe the other group members know better.

Photo 13.1. A group picture of us in 1991. Above from left to right: Ahmed, Peijun Cong, Earl Potter, and Ian Sims. Middle from left to right: Dean Willberg, Ahmed Heikal, Luis Bañares, Bob Bowman, and me. Bottom from left to right: Chuck Williamson, Marcos Dantus, Jennifer Herek, and Martin Gruebele.

Michael Gutmann received a diploma degree in Chemistry from the University of Cologne in 1985. He did his graduate work in two-photon laser spectroscopy at the University of Cologne in the laboratory of Georg Hohlneicher and received a Ph.D. in 1988. From 1990 to 1992 he was a postdoctoral researcher under Ahmed H. Zewail at the California Institute of Technology. He then returned to the University of Cologne as a research fellow and established his own research group. He finished habilitation (tenure track) in 2000. His main research areas were in femtochemistry of molecules and clusters, in particular metal carbonyl and solvent clusters.

In 2000, Michael left academia and worked for a consulting company. In 2001, he joined the engineering company Nienstedt GmbH as Senior Project Manager and became Vice President in 2007. In this position, he focused on international sales (Europe, Asia, and USA) and marketing.

In 2012, he founded LIOP-TEC GmbH together with his former graduate student, Jürgen Lindener-Roenneke. LIOP-TEC GmbH develops, manufactures, and distributes high quality opto-mechanical components and tunable nanosecond dye laser systems and accessories. Customers are mainly scientists and the laser industry. Michael is co-owner of LIOP-TEC GmbH and serves as managing director.

Michael is married and has two sons.

14 Insights on Molecular Structure and Real Time Dynamics: An Energy Flow Inspiration by Ahmed Zewail

Luis Bañares*

I first met Ahmed in 1989 when he came to Spain for the first time to attend and deliver two lectures at a summer school at El Escorial, close to Madrid, under the Summer Schools of University Complutense of Madrid. Ahmed was invited by my Ph.D. supervisor, Ángel González Ureña, and I was the secretary of the school. I was in charge of taking care of Ahmed, so I went to Madrid Barajas airport to receive him and took him to El Escorial in my second-hand old-fashioned Renault 7. He did not complain though. He spent a week in El Escorial and delivered two exceptional lectures in front of mostly Spanish students. It was really shocking for a Spanish Ph.D. student like me, working in crossed-molecular beam experiments, to be able to listen to Ahmed present the latest experimental findings in Femtosecond Transition State Spectroscopy (later named femtochemistry) carried out at Caltech using femtosecond lasers. I was so impressed that I decided to do my postdoctoral research with him. On the way back to the airport, Ahmed invited me for lunch and then I expressed my desire to do my

* Postdoctoral Research Fellow/Visiting Associate.
Email: lbanares@ucm.es

postdoctoral research with him starting around September 1990 once I had finished my Ph.D. I got a positive answer from him and about a year later, I was traveling to Pasadena.

When I started my postdoctoral research with Ahmed, he proposed that I work in the 036 laboratory together with an Egyptian Ph.D. student, Ahmed Heikal. The first project was related to studying the vibrational predissociation in anthracene-argon van der Waals complexes using picosecond time-resolved spectroscopy with time-correlated single-photon counting. This technique had been used successfully by others in the group, including Felker, Baskin, and Semmes, in impressive applications such as quantum beats or rotational coherence in polyatomic molecules. The project ran very well and we worked really hard to get results. By February 1991, we had finished the experiments and we wrote a paper that was later published in *Chemical Physics*, volume 156 (issue 2, pages 231 to 250). During this time, I learnt a lot from Ahmed not only in performing high-class experiments, but also in doing good science.

In March 1991, when we wanted to carry on new experiments in 036, we ran into some bad luck when the Ar^+ ion laser tube broke down. Ahmed was in a long trip to Europe and he did not give us his approval to repair the laser. He told us that we should wait until he was back in Pasadena and we should instead study and make proposals for a new project in 036 in the mean time. Ahmed Heikal and I took this seriously and we were studying and reading the literature in search for a new project as Ahmed had proposed. This took some three weeks in total from the end of April to mid-May 1991. Then in the evening of 18 May 1991, Ahmed told us to meet in an Egyptian Café at Hollywood Boulevard in Los Angeles. It was an amazing meeting with Turkish coffee and shisha. Once there, Ahmed Heikal and I started to tell him what we had found for a new project in 036, and Ahmed listened silently but carefully as we presented all our proposals. When we finished, he said all of a sudden in his typical enthusiastic way, "All these proposals seem very interesting and you did a great job; however, this is what you are going to do." And then he took the envelope where I had put the proposal sheets and started to make drawings while he was talking. All the drawings Ahmed did

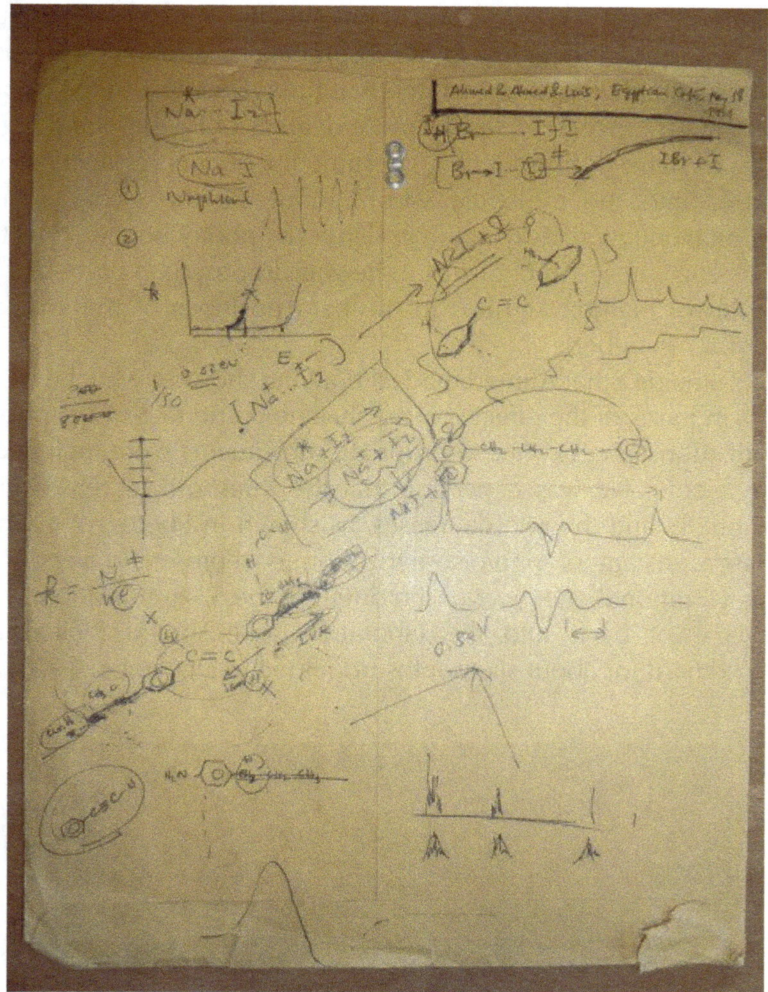

Photo 14.1. Picture of the envelope with Ahmed's drawings to explain his proposal of experiments in 036. These drawings were done during our meeting at an Egyptian Café in Hollywood Boulevard (Los Angeles) in the evening of May 18, 1991.

are in Photo 14.1, but I want to emphasise two particular drawings that were especially relevant and enlightening for the next science to be carried out in 036.

The first drawing is in Photo 14.2 and represents Ahmed's ideas about energy flow and dynamics in the photoisomerization of stilbenes. He was very interested in investigating molecular structural effects, based on the inclusion of substituents in different positions of the trans-stilbene molecule, on the real time dynamics of the photoisomerization process. Including more degrees of freedom and higher density of states by introducing substituents, donors, or acceptors of electrons into the molecule would influence the intramolecular vibrational energy redistribution and thus the photochemical reaction rate. As you can see in Photo 14.2, he was proposing in particular the inclusion of methoxy groups in the para positions of the aromatic rings of trans-stilbene.

It is amazing that Ahmed made a drawing showing what he would expect in terms of the effects on the threshold and rate of the photoisomerization process once appropriate substituents were included in the molecule. He was expecting that the threshold increases quite substantially and the rate decreases, as shown in Photo 14.3, where the rate constant of photoisomerization is represented versus the excess vibrational energy over the origin of the S_0–S_1 transition.

We went back into the laboratory and Ahmed Heikal and I worked hard for about six months from June to December 1991. We

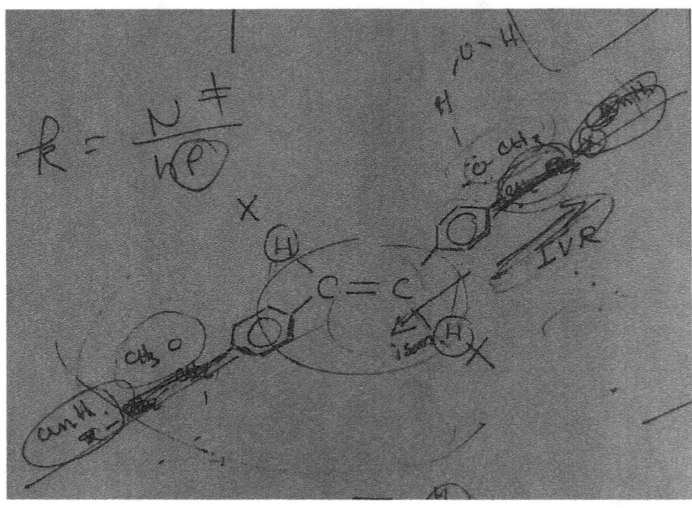

Photo 14.2. Magnification of the drawing where Ahmed illustrated the idea of energy flow and structural effects on dynamics by including substituents in the trans-stilbene molecule.

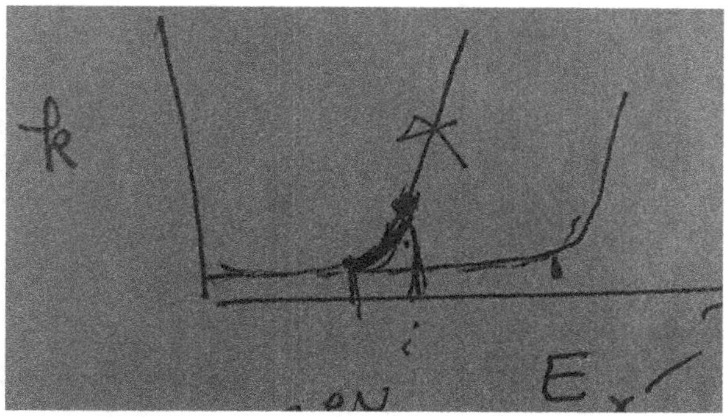

Photo 14.3. Ahmed's predictions for the threshold and rate of the photoisomerization of substituted trans-stilbenes.

studied several different molecules: 4-methoxy-trans-stilbene, 4,4′-dimethoxy-trans-stilbene, 4-hydroxy-trans-stilbene, and 2-phenyl-indene. The idea was to see the effect of the structural modifications on the potential energy surfaces involved, on the density of states, and on the intramolecular vibrational energy redistribution, and in turn on the photoisomerization threshold and rate. The results were amazingly consistent with Ahmed's expectations (see Photo 14.4)! This is an example of Ahmed's impressive intuition in science.

This work was published as a communication in the *Journal of Physical Chemistry* in 1992, volume 96 (issue 11, pages 4127 to 4130) and later in a full paper published in the same journal in 1997, volume 101 (issue 4, pages 572 to 590), with the participation of Spencer Baskin who joined the group as a senior scientist after I left Caltech in September 1992.

Now we jump ahead about 15 years to the summer of 2007. I was spending a sabbatical at Caltech and Ahmed wanted me to collaborate with a young Ph.D. student, I-Ren Lee, to study the photodynamics of the trans-stilbene anion! Ahmed always loved this molecule and he was really keen to dedicate more effort to its study, in this case using time-resolved photodetachment and photoelectron spectroscopy. I-Ren acquired some nice results and we published a communication in the *Journal of the American Chemical Society* in 2008. During my time at Caltech that summer, I gave a couple of group seminars. In particular, one of my lectures was related to the recent results we were

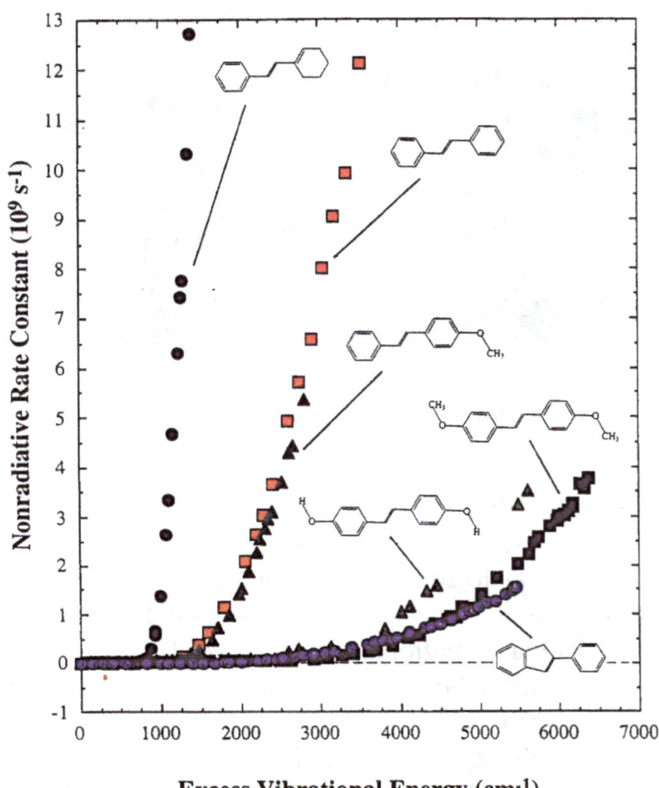

Photo 14.4. Photoisomerization rate constant as a function of the excess vibrational energy for trans-stilbene and all the derivatives and substituted molecules studied. Modified from *Journal of Physical Chemistry A*, volume 101 (issue 4, pages 572 to 590).

getting in Madrid on the real time dynamics of the photodissociation of a polyatomic molecule, methyl iodide, by combining femtosecond pump-probe, resonance-enhanced multiphoton ionisation detection of the fragments, and velocity map imaging. After spending many difficult years trying to get funding in Spain, by 2004 or so we were able to start researching on femtochemistry! After my talk, Ahmed and I went for a coffee and then he came up with another original idea: looking for structural effects (again energy flow!) in the real time dynamics of the alkyl iodide family of molecules. The idea was to see the effect of the structure of the molecule on the energy flow and therefore on the reaction times; that is, what would be the role of the molecular structure on the real time dynamics for a family of molecules sharing the same reaction coordinate? His research question and proposal was very timely in terms of the work we were carrying out in Madrid. However, it took a while to measure (through clocking experiments) all the molecules: methyl, ethyl, *n*-propyl, and *n*-butyl iodide as linear molecules, and *i*-propyl and *t*-butyl iodide as branched molecules (see Photo 14.5).

Ahmed came to Madrid in May 2008 when he was awarded an honorary degree from Universidad Complutense of Madrid. I had the honour of being his Godfather as I presented him to the University community in a very nice ceremony. This was a special occasion that he always remembered affectionately. By July 2011, we had plenty of results and during the celebration of Femto10 (the Madrid Conference on Femtochemistry), Ahmed and my group had several exciting discussions (see Photo 14.6). Again, Ahmed's intuition played a key role in the interpretation of the results. Later, I met Ahmed at the Ritz-Carlton Hotel in Berlin on 8 March 2013 and we spent several hours discussing and writing a draft of a manuscript. I can remember very well that he was complaining of a strong pain in the back. We

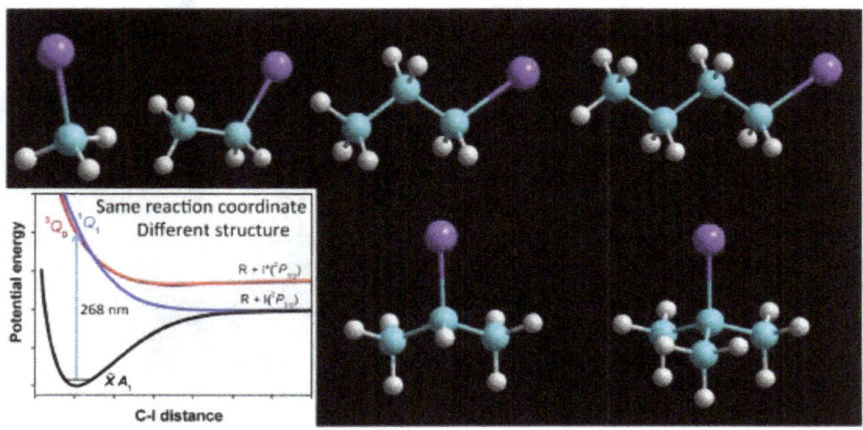

Photo 14.5. Series of alkyl iodide molecules whose real time photodissociation was studied by femtosecond pump-probe velocity map imaging to disentangle the role of the chemical structure on the energy flow and real time reaction dynamics.

Photo 14.6. Discussion in the Faculty of Odontology, University Complutense of Madrid, during the celebration of Femto10, The Madrid Conference on Femtochemistry, in July 2011.

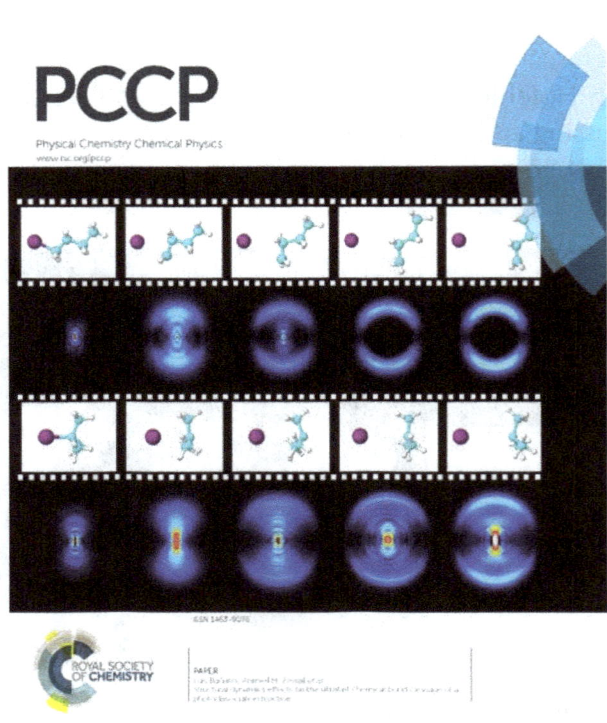

Photo 14.7. Cover of issue 19 in volume 16 of *Physical Chemistry Chemical Physics* highlighting our article, published in 2014.

Photo 14.8. Ahmed and me at Alqahira restaurant in downtown Toledo, Spain, on July 23, 2015.

finally published the work in 2014 in *Physical Chemistry Chemical Physics*, volume 16 (issue 19, page 8703) and we got on the cover of the issue (see Photo 14.7)!

Ahmed came to Spain for the last time in July 2015 when he gave the inaugural plenary lecture in the International Conference on Photonic, Electronic and Atomic Collisions organised in Toledo. He was in really good shape! I enjoyed his lecture and his enthusiasm and we spent a few days together in a city that he specially loved. I remember having dinner in the Parador and in several cigarrales and specially the lunch at an Egyptian restaurant named Alqahira (which means Cairo in Arabic) in downtown Toledo, the city of the three cultures. I would like to end off with a picture that Dema took of the two of us in Alqahira (see Photo 14.8) where you can see Ahmed smiling and enjoying life. Thanks Ahmed for all what I have learnt from you! I will miss you!

Luis Bañares received his Ph.D. in Chemistry from Universidad Complutense de Madrid (UCM) in 1990. Following postdoctoral research associate positions at California Institute of Technology (1990–1992) and Universität Würzburg (1995–1996) with Fulbright and Alexander von Humboldt fellowships, respectively, he joined UCM first as an assistant professor followed by associate professor, and since 2007 he is a full professor at the Department of Physical Chemistry of UCM. Since 2014, he is the director of the Center for Ultrafast Lasers at UCM. He is Associate Editor of the *Physical Chemistry Chemical Physics* journal and fellow of the Royal Society of Chemistry. His research interests are related to experimental and theoretical chemical reaction dynamics and femtochemistry. His work focuses on the understanding and strong laser field control of fundamental chemical reactions and photodissociation processes at a molecular level.

15 In Memory of Dr. Ahmed H. Zewail

Charlie Qianli Liu*

I joined Dr. Zewail's group in the summer of 1991. It was at a time when I entered my second year at Caltech and had the opportunity to select a new thesis advisor. Dr. Zewail was well known on campus for his pioneering work in applying ultrafast lasers to the study of molecular reactions. Before the meeting with Dr. Zewail, I was quite nervous as I did not know what to expect in the conversation, but he received me in his office with a warm smile. He looked young, energetic, and considerate, and he was particularly interested in me when he learned that I majored in physics and that I had passed all my written candidacy requirements for the Ph.D. program in my first year. He called up my previous advisor and confirmed with me that I could join his group right away.

Dr. Zewail took me to his laboratories in the sub-basement of the Noyes building. There were four of them and we stepped into one that was very noisy and full of laser lights. Mr. Earl Potter was working there and Dr. Zewail introduced us and told us that we would work together in that laboratory from then on.

Dr. Zewail had chosen the research focus for the laboratory. The experiments were to examine how solvent affected the behavior and dynamics of the Iodine molecule (I_2). We chose helium, neon, or argon gas at various pressures as a solvent. The sample system ranged

* Graduate Research Assistant — Ph.D.
Email: Charlie@vxichina.com

from Iodine and solvent in high-pressure cells to Iodine-solvent gas mixtures injected to a vacuum chamber. The laboratory setup was to use one laser pulse ("pump pulse") to excite I_2 molecules to an intermediary state, and then use a second pulse ("probe pulse") to push the molecules to an even higher excited state that can subsequently emit a signature fluorescent light. By monitoring that fluorescent signal while changing the relative delay time between the pump and the probe, we would be able to track the excitation process of I_2 molecules in real time.

The idea sounded simple, but the laboratory setup was quite complex. Earl had spent years building and assembling the instruments and apparatus required. I jumped in at a perfect time to get on the experimental work. The biggest challenge was to get stable laser pulses at the right frequency and to make the pulse duration short enough. The commercial lasers available then could only offer laser pulses at a fixed wavelength and only with pulse durations longer than 10^{-12} seconds. We needed to get the pulse duration down to 10^{-13} seconds or less. The wavelengths of the pulses also needed to be tuned to that of the resonant wavelengths of Iodine molecules. To achieve that, a home-built dye laser, a multi-stage dye amplifier, pulse compression prism pairs, and frequency-doubling devices all needed to be perfectly aligned and tuned. It typically took hours just to get both the pump and probe beams aligned. We then needed to cross-check the pulse duration. If they were longer than what was required, we would need to adjust the dye laser, dye amplifiers, and prism pairs once again. Quite often when everything appeared to be in good shape and we managed to focus the beams properly into the vacuum chamber, something would just change even slightly and mess everything up. It could be something as small as a minor change of reflection angle of a spring-loaded mirror or a slight decay of the dye performance in one of the dye cells due to laser exposure. For weeks, we worked daily from early morning til late at night without getting any meaningful results. Dr. Zewail came down to the laboratory almost every day. He was apparently very familiar with the laboratory setup. When I got exhausted trying a whole day without getting anywhere, Earl comforted me by saying that Dr. Zewail used to work like us in the laboratory for many years.

I was quite fortunate that the hard work paid off after a few months. One night around 8.00 p.m., Earl and I completed the experimental setup and everything went well that day. We started the computer-controlled data collection program and went outside the laboratory to take a breather, sitting on the floor along the corridor and chatting for a short while. We then went back into the laboratory to check on the results and to our surprise, the computer screen showed a data trajectory that was amazing and unexpected. We quickly checked through the entire system setup, recalibrated the laser pulses, and then successfully replicated the result. Earl called Dr. Zewail and he came down right away from his office. At that time, we did not know exactly what the finding meant yet, but the experimental results were real, promising, and exciting.

During the following weeks, we ran the experiments repeatedly while changing sample conditions and pump and probe pulse frequencies systematically. More and more results were obtained following the initial observation. What we discovered in the laboratory was apparently related to phase correlation and coherence dispersion of molecular vibration in solvent-like environment. Some of the experiments also demonstrated what we called the "caging" effect. Dr. Zewail instructed us to prepare laboratory reports and research articles, and I was asked to draft one of them. It was the first time I wrote a formal article in English. With quite some effort, I finished the draft and sent the manuscript to Dr. Zewail. He called me to his office shortly after and pointed out hundreds of grammar, spelling, and punctuation errors. He also asked me to check and verify that all references were quoted and indicated properly. I redid the entire manuscript, and once again Dr. Zewail pointed out dozens of issues and errors. This was followed by a third and fourth time, and errors were still found in my drafts. At last, at the fifth round of editing, he said he could let go of the paper, but he told me that correcting my writing errors was the most painful and time consuming thing ever. He expected me to learn from this process and check my own writing carefully in future before submitting to others.

Earl graduated about a year later with a Ph.D. degree in Chemistry, even though he joked sometimes that his research had little to do with chemistry. He certainly contributed tremendously to the setup and success of the laboratory. I continued to run the laboratory on my own for several months before Dr. Juen-Kai Wang joined. Juen-Kai came with professional experimental experience and excellent academic training. Working together with him, we continued our experimental series and produced more and more significant and interesting results over the next two years.

While continuing with the laboratory work, I was frequently called to Dr. Zewail's office to discuss the interpretation and implications of our experimental results. Dr. Zewail expected me to be able to explain our observations in detail and using mathematical formulas. Knowing the principles of physics was one thing, but formularizing the coherence behavior of molecular vibration in complex systems was another. I had to do tons of readings to prepare for the discussions with him. Dr. Zewail is genuinely interested in science and I have not seen anyone else, until today, who is more committed to exploring the fundamentals behind the laboratory results. Dr. Zewail would not accept anything "roughly," "generic," or "in principle." He expected clear and direct answers. In particular, he expected quantified answers or mathematical proofs that could be derived from A to Z. His questions often got me scratching my head. Eventually, I proposed to Dr. Zewail that I would design and create a computer simulation program, thereby putting theories, assumptions, and experimental conditions all together to reproduce the experimental results.

Dr. Zewail fully supported me. He quickly procured the required workstation and arranged for sufficient computer time for me at a supercomputing center located in San Diego. I started writing computer codes that simulated the dynamics of I_2 molecules inside Helium, Neon, or Argon clusters of various sizes. It took me several months to get the first version of the simulation program done. There were about a million lines of codes written during that period of time. Running the program at the Cray supercomputers located in San Diego, I found that it would take a very long time to compute the dynamics of cluster systems that included more than 100 particles, not to mention that the ensemble average of thousands of such systems was needed to simulate a real life situation. Besides improving the code efficiency, I needed to find faster computing resources. Luckily for me, there were such resources. At that time, Intel had donated to Caltech a state-of-art parallel computer system called Alpha. It came with 536 nodes of CPU, each with supercomputing power. Unlike the Cray supercomputer at San Diego, Alpha's available CPU time was plenty. To leverage the multiple nodes of Alpha, programs needed to be adjusted and compiled for parallel processing. At that time, very few people knew how to do that, and neither did I initially. I talked to Alpha's system administrator a few times. He was extremely nice and supportive and he encouraged me to try figuring it out. He said he felt sorry that Alpha was not being utilized effectively. After several times of trying, I found a trick and an algorithm to run my program on Alpha's multiple nodes. Eventually, I was able to run my simulation program simultaneously on all 536 nodes, each computing a statistically independent trajectory. The CPU usage was so high on each of the nodes loaded that the system administrator called me with excitement and asked me how I did it. I shared with him the idea and he encouraged me to use as many nodes and as many CPU hours as I needed. For the next two to three years, I probably used more of Alpha's CPU time than all the others combined. That sharply reduced the time required to complete my series of simulation work. The work eventually reproduced most of the laboratory findings. Dr. Zewail followed up closely with me on virtually every step of the computer simulation process. Whenever there was progress, he got excited as much as I did, which gave me the extra motivation to constantly work on the improvement of models and computation algorithms throughout the years.

Before joining Dr. Zewail's group, I was not good at giving speeches in front of an audience. At that time, my English proficiency was barely enough for a normal conversation. We typically had group meetings once or twice a week and almost every time, Dr. Zewail would ask me to present something in front of the group. I somehow became more and more confident over time and my verbal English also improved significantly. One year later, when I presented my oral candidacy materials in front of my thesis committee, Dr. Tombrello, who had been my first-year advisor, was so shocked to notice my change and progress. Over the last twenty years after graduating from Caltech, I have made countless speeches often in front of a big audience and without prepared scripts. I believe that I would not have been able to do that without my early experience working for Dr. Zewail.

Dr. Zewail's research group attracted the most talented students and young scientists from all over the world. The group had about a dozen members divided into four laboratories and two offices. Every laboratory was doing unique and amazing research work and people in the group were all open and supportive of each other. We all anticipated that Dr. Zewail would one day be awarded the Nobel Prize. I thought the award would come before my graduation, but it waited until two years later.

Besides the science talks, I had other opportunities to get to know Dr. Zewail outside of research work. Dr. Zewail invited me to dinner on many occasions. At one point, I suggested to Dr. Zewail that he should do more exercise to stay healthy, and he listened. For a long while, we played tennis together once or twice each week. I was not good at tennis, but I enjoyed the time very much. As a Chinese student studying abroad, I was lucky to have his warmth, care, and friendship. Outside campus, we often talked about life, culture, and social issues in the US, China, and Egypt. Dr. Zewail was very interested in knowing more about the political and social changes happening in China. He mentioned several times that he hoped for similar economic improvements in Egypt.

Combining the results of the experimental studies and computer simulations, I was able to complete my Ph.D. program in the summer of 1996. I missed the graduation ceremony of 1996, so I waited until 1997 to get my official degree award. Dr. Zewail had enquired about my post-graduation career plans and he strongly recommended that I should continue with academic pursuits. When he learned that I would take a very different path, he was quite surprised. However, he eventually became supportive and also expected me to do well in the commercial world.

I went back to see Dr. Zewail in his office on 1 December 2000. That was one year after I went back to China and before I began my third startup venture. He was very happy to see me again and he had reserved an official copy of the publication from the Nobel Prize Committee for me. The publication captured the moment when he was awarded the Nobel Prize at the world's most renowned celebration of science. He wrote the following and signed his name: "To dear Qianli, one of my great students with best wishes as always! Ahmed Zewail." I was deeply moved.

Many years passed. Every now and then, I thought about going back to see Dr. Zewail. Somehow, I felt as if I had not lived up to his expectations. I wanted to achieve more so that I could share more of these moments of success with Dr. Zewail when I met with him again and perhaps come within range of his class. I was wrong. In recent years as I grew older, I realized that I could not be nearly as successful as Dr. Zewail no matter what I do. In fact, very few people on this planet could have achieved nearly as much as he did. During the summer of 2016, pieces of old memories often came to my mind. I had a strong feeling that I should go visit Dr. Zewail soon. Within days, I was saddened by the tragic news that Dr. Zewail had left this world.

Dr. Ahmed H. Zewail, my advisor, my mentor, and my dearest friend, had probably embarked on another wonderful journey in another nice world. My best wishes to him, as always!

Charlie Qianli Liu enrolled in the physics department of California Institute of Technology as a graduate student in 1990. He completed his thesis work with Dr. Ahmed H. Zewail's guidance in the summer of 1996 and was awarded his Ph.D. degree the following year. Prior to that, Qianli graduated from Beijing University in 1988 with a B.S. degree in physics.

After graduating from Caltech, Qianli joined International Communication Enterprise, a startup company specializing in software design and development. In 1998, Qianli joined KillerBiz.com as Vice President of research and development, leading the design and development of websites and e-commerce applications.

Qianli returned to China in 2000 and joined JiaJia123.com as a co-founder and Chief Technology Officer. One year later, Qianli joined VXI Global Solutions and established its China operations, known as Vision-X Enterprise Management Ltd (VXI China). VXI China focused on the development of contact center technologies and outsourced contact center services. As President and Chief Executive Officer, Qianli led the growth of VXI China from a few employees to its present-day strength of 10,000 employees. VXI China is now one of the leading companies in China's contact center industry.

16 Where Science, There Sunshine

Thomas Baumert*

My first encounter with Ahmed was at a spring meeting of the German Physical Society in 1991 which was held at Universitaet Freiburg, Germany. Ahmed gave an invited talk and visited our laboratories afterwards. At that time, I was approaching the end of my Ph.D. studies with Gustav Gerber. Ahmed was fond of the amplified colliding pulse mode-locked laser system that I built during my Ph.D. and especially our experiments devoted to femtosecond pump-probe experiments to study the dynamics of multiphoton ionization and fragmentation of molecules and small metal clusters. So he invited me for a beer in front of the famous Freiburg Minster, and in the afternoon sunshine he talked about research possibilities at Caltech and the beauty of California.

Back then, I was still undecided about whether I should proceed with an academic career, but that afternoon turned out to have a strong influence on my future. A little bit flattered and very much tempted by the prospects of doing science at an even more sunny place than Freiburg, I started to write a grant proposal to the German Research Foundation asking for money to do time-resolved molecular dynamics on surfaces. The grant proposal got accepted and so I travelled in the summer of 1992 to Caltech. I arrived just in time on a Friday afternoon for the weekly group meeting with an extremely

* Postdoctoral Research Fellow.
Email: tbaumert@physik.uni-kassel.de

warm reception by Ahmed and the other members of the group. Michael Gutmann took care of me and helped me get settled and after the weekend, I was the owner of an old police car and had rented a flat. On Monday morning when I met Ahmed, he kindly enquired about my holiday, so I told him about the beauty of Crete where I spent a week with my fiancé before coming to the States. In a friendly manner, he listened to my long holiday report and at the end he said with his special good humor, "Thomas I was actually referring to the weekend," and we both laughed heartily.

The following year at Caltech was extremely productive and, together with Jennifer Herek and Soeren Pedersen, we wrote three papers. During that time, I had the chance to interact a lot with Ahmed in the laboratory and discuss results in his office. I enjoyed his straight style of doing science which always involved going back to first principles. The barbecue invitations to his home where he was the perfect host are also memorable events. When the year was over, Ahmed extended an offer to me to continue working in his laboratories. However, as I wanted to get married and proceed with my career in Germany, I had to decline his generous offer. After my wedding on 4 July 1993 in Las Vegas, I left the States and went back to Universitaet Freiburg. After my Habilitation at Universitaet Wuerzburg and a position as head of the LIDAR group (light detection and ranging) within the German Space Agency, I am since 1999 a full professor of experimental physics at Universitaet Kassel.

Notable encounters with Ahmed after my postdoctoral period were the "Lasers 98" in Arizona, where part of the Zewail family met together in a hotel suite rented by Marcus Motzkus and we enjoyed cigars and good conversation. The reception in the Egypt embassy during the 220th American Chemical Society Meeting in Washington where a Nobel symposium was held for him was impressive. Later in 2005, I had the pleasure of hosting him for two days at the German Physical Society spring meeting in Berlin where he gave a plenary lecture. I was very much impressed that despite his many duties he still enjoyed hours of detailed scientific discussions, in particular on "photon locking," a topic that he had investigated in the 1980s and we restarted to investigate with more advanced laser methods being available in the meantime. My last personal encounter with Ahmed was in 2013 at the 125 Years Anniversary of Angewandte Chemie in Berlin. There, he literally had to be carried away from the podium after an inspiring talk because of serious back pain. We had only seen each other briefly and exchanged a few words, but despite his pain he was sincerely delighted to see me again. I am grateful for that spring afternoon in front of the Freiburg Minster and the attitude to life "where science, there sunshine" that Ahmed exemplified.

Thomas Baumert was born in 1962 and studied physics at the University of Freiburg. He obtained his Ph.D. which was devoted to femtosecond spectroscopy of molecules and clusters with distinction in 1992 and was awarded the Gödecke prize. A postdoctoral research fellowship from the German Research Foundation (DFG) allowed him to join Professor Ahmed Zewail's group at Caltech from 1992 to 1993. He finished his Habilitation in 1997 at the University of Würzburg in Professor Gustav Gerber's group with the help of a DFG Habilitation scholarship as well as support by FCI (Fonds der Chemischen Industrie). He was also awarded a Heisenberg scholarship under the DFG and joined the German Space Agency in Oberpfaffenhofen where he was head of the LIDAR group until 1999. He then accepted a full professor position for experimental physics at the University of Kassel. In 2000, he received together with Professor Gustav Gerber and Dr. Volker Seyfried the Philip-Morris Science Award for research on the control of chemical reactions by femtosecond laser pulses. His research specialization is in femtosecond spectroscopy and ultrafast laser control, the latter in relation to quantum control, control of chemical reactions and the development of fs-laser-based techniques for nonlinear microscopy, spectroscopy and nanostructuring. Recently, he started a new research direction in ultrafast electron diffraction.

17 Working as a Postdoc at Caltech

Christoph Lienau*

My voyage to Femtoland started early in January 1992. I was just about to finish my Ph.D. in physical chemistry under Professor Jürgen Troe at the University of Göttingen. During my thesis writing, I was trying to look at the effects of solute-solvent interactions on the photoisomerisation dynamics of stilbene, a popular topic at the time. My task was to build a molecular beam apparatus and to perform picosecond time-resolved fluorescence spectroscopy on isolated stilbene molecules and stilbene/hexane van der Waals complexes in supersonic jets at low temperature. I repeated some of the groundbreaking spectroscopy work performed a few years earlier by Peter Felker and Jack Syage from Professor Zewail's group. I was intrigued to see the first clear signs of rotational recurrences in my own data and started studying the quantum dynamics of rotational wavepackets by reading Zewail's paper. I gladly finished my thesis by measuring rotational recurrences in stilbene/hexane clusters with a deep respect for what I had learned about quantum coherences from Zewail's work. All my experiments were just picoseconds in time resolution, limited simply by the single photon counting system that I had built and thus far away from truly ultrafast experiments.

* Postdoctoral Research Fellow.
Email: christoph.lienau@uni-oldenburg.de

At the time, however, some of my fellow Ph.D. students, Dirk Schwarzer and Peter Vöhringer, already started the first femtosecond pump-probe experiments in Göttingen in a team led by Jörg Schröder. We were of course closely following what the famous group at Caltech was doing. We were fascinated by all the spectacular experiments probing the motion of coherent vibrational wavepackets in iodine as well as during the photodissociation of sodium iodide and struggled hard to understand the underlying quantum mechanics. Hence, with the benefit of hindsight, it might seem not too surprising that I sent an application letter for a postdoctoral position to Ahmed. It was the only application letter that I ever wrote for such a position. In this letter, I tried quite hard to convince Ahmed that I am a talented young physicist who would love to work in his group. A little more than a week later, I got a nice letter back from Caltech. Ahmed replied saying that my supervisor is an excellent colleague and good friend of his and that he would be more than happy to accept me as a member of his group. Jürgen then advised me to apply for a postdoctoral fellowship from the German Research Foundation (DFG). I immediately sat down and wrote, in just two days, my first ever DFG proposal. Of course, I was not yet aware of the high scientific standards that are required to get such proposals accepted; otherwise, I would certainly have spent more time on the writing. In my proposal, I suggested studying the predissociation of iodine in high-pressure, supercritical rare gases, at densities that bridged the gap between the gas and liquid phases.

Jürgen advised me to be optimistic about the proposal application and generously offered 20,000 German Marks as immediate travel support. This, together with some additional financial support by Ahmed, allowed me to start on my new work even before a decision about my DFG application was reached. Shortly before that, I had the opportunity to meet Ahmed in person for the first time. I attended the ultrafast phenomena conference in Antibes where Ahmed gave an invited talk and introduced the ultrafast electron diffraction project he had just started to work on. I was fascinated by the idea of "simply" using an ultrashort electron bunch to obtain transient structural information about molecules. More importantly, I had the opportunity to touch base with such an inspiring personality. It was immediately clear that Ahmed truly loved science and had a clear vision of what he wanted to achieve. Back home in Göttingen, I convinced my wife Christiane to leave Germany for a while. So, in July 1992, we grabbed our two children, Theresa and Jakob (33 and 19 months old at the time), packed five suitcases and a bicycle, and flew to Los Angeles. Thomas Baumert picked us up from the airport and had a hard time fitting all of us and our luggages into his Buick Sedan. Thomas was already famous in Germany at the time for the beautiful doctoral work he did on sodium clusters that was performed in Gustav Gerber's group, and it was a pleasure to meet him in person. Coming from cosy little Göttingen and being in the US for the first time, the ensuing car ride from the airport to Pasadena was quite spectacular.

Very soon, I got introduced to the work that was going on in the basement of the Noyes laboratory at Caltech. Ahmed gave me a few days to settle in before he asked me to start working on a project together with Ahmed Heikal. We were asked to study vibrational relaxation phenomena in stilbene/hexane van der Waals clusters, a follow-up of my Ph.D. work. Ahmed was a very friendly and kind person and already had a lot of experience with the molecular beam system in the laboratory. We formed a good team and quite quickly found pronounced quantum beats in the time-resolved fluorescence of such clusters. Jointly, we wrote a paper comparing intramolecular vibrational-energy redistribution in isolated stilbene molecules and stilbene/hexane clusters. Ahmed Zewail introduced us to the art of scientific writing and I had the pleasure of participating in a few after-dinner meetings where we meticulously went through every single line of the manuscript until well after midnight. The paper quickly got accepted in *Chemical Physics*. Later, I learned that it was actually a big mistake to send this paper out so early. It got published well before some of the results of my Ph.D. thesis were submitted for publication and my supervisors in Göttingen were of course very unhappy to see it in print. I deeply apologize to them for this naïve mistake.

The intensity of the work that I experienced in the writing sessions with Ahmed was present throughout the Noyes laboratories. Before I came to Caltech, I thought that I was self-motivated and eager to learn new things. I also thought that I knew how to work hard and get my experiments going. I soon realized that different work ethics were required to survive at Caltech. I got used to the after-dinner laboratory work which was often interrupted by phone calls from Ahmed asking about our progress while he was on travel. I also learned that it was a good idea to run some experiments over the weekends, something that we "German postdocs" (at the time there were three of us in the group — Thomas, Arnulf Materny, and myself) and our families did not always appreciate. Since everyone in the group worked quite hard (at least in my opinion), we quickly adapted. The enormously stimulating atmosphere at Caltech, with its huge number of extraordinarily talented Ph.D. students and postdoctoral researchers from all over the world, made it easy to adapt. Working during the weekends also sometimes gave me the opportunity to talk to Ahmed in a more relaxed atmosphere. As everyone knows, Ahmed was an extraordinarily charming and pleasant person to talk to and I greatly enjoyed these rare occasions of private communication.

In parallel to the stilbene work, I started to prepare the design of the high-pressure cell that I needed for the iodine experiments I had proposed to the DFG. My DFG fellowship had been granted and I was eager to start the work. So, I called Ahmed in his office and explained to him that I had finished the design of the experiment and that I knew exactly how to build the required high-pressure cell. The necessary equipment could be provided by a manufacturer in Switzerland and would just cost a mere US$30,000. Ahmed immediately hung up the phone, obviously shocked by the enormously expensive equipment that I requested. A few hours later, he called me back and asked me to go ahead with ordering the equipment.

A few months later, I finished the construction of the high-pressure equipment. The cell withstood all pressure tests and we could start femtosecond work. Marcos Dantus kindly introduced me into the secrets of colliding-pulse mode-locked lasers and dye amplifiers. So, we generated 50-fs pulses and sent them into the cell — out came a beautiful and colorful conical supercontinuum emission. All colors of the rainbow were sent into a conical beam with a large opening angle.

Without knowing much about supercontinuum generation yet, I was simply disappointed that the planned experiments did not work out and was afraid that my entire DFG proposal would be a total failure. Luckily, by replacing the sapphire windows of the cell with quartz windows, supercontinuum generation could be suppressed and the optical experiments could be started. I vividly recall the moment when I saw the first oscillations in the pump-probe signals due to the coherent vibrational wavepacket motion of iodine in high-pressure gases. We could follow the effect of the rapid random collisions of the iodine molecules with the buffer gas atoms on the vibrational motion and could see how these collisions led to the dissociation of the iodine molecule. At sufficiently high pressures, the data nicely showed the geminate recombination of the dissociated iodine molecules inside the cage of solute molecules. Ahmed was very excited about the results and asked me to document the first set of data in a paper for *Chemical Physical Letters*. After two or three sessions of after-dinner writing, the paper went out and immediately got accepted.

Ahmed then asked me what I wanted to do next. I tried to convince him to systematically study the effects of different buffer gases on the dynamics of iodine, exactly as I had proposed it to the DFG. Ahmed halfheartedly agreed and teamed me with Arnulf Materny. Arnulf and I set out to perform a long series of measurements, testing numerous buffer gases and varying the buffer pressure in small steps. All these experiments provided interesting results but essentially showed the same physics — iodine coherent vibrational wavepacket motion, predissociation, and recombination — as before. Since we had lots of data, Arnulf and I needed a long time to analyze them. In particular, I tried very hard to theoretically analyze the interesting coherent vibrational energy relaxation dynamics that was contained in the data. Unfortunately, I failed since I did not know enough yet about Redfield and Lindblad equations to analyze the data properly. Consequently, and not surprisingly, Ahmed lost a bit of interest in this project. I learned that the proverbial German thoroughness and American excitement for scientific breakthrough can be at odds and I perhaps missed some opportunities to work on other exciting projects in the group. Nevertheless, we ended up publishing four more very nice papers together and, with great help from Ahmed, the project was quite successful in the end. Most importantly, I not only learned a lot about the technical details of scientific work from Ahmed but also greatly admire his passion and enthusiasm for science.

After 20 wonderful months at Caltech, my family and I went back home to Germany and I took a position as a research scientist at the newly founded Max-Born-Institute in Berlin. Here, I joined Thomas Elsässer's department and initiated activities that were at first only remotely connected to my work with Ahmed. With Thomas, I was trying to combine ultrafast spectroscopy and the newly emerging techniques of near-field optical microscopy to probe the motion of electrons in semiconductor nanostructures. For quite some years, we went deep into the field of coherent semiconductor optics and focused on studies of the optical response of single nanostructures. Hence, we almost forgot about all the exciting work at Caltech, in particular in the field of ultrafast electron diffraction and microscopy. Somewhat later, we became interested in plasmonics. Based on this work, Claus Ropers and I started some ten years back to focus few cycle light pulses to the apex of sharp, nanometer-sized gold tips, discovering that this transforms these tips into ultrafast, nanometer-sized electron guns with a tremendous potential in ultrafast electron microscopy. Two years back, my coworkers Jan Vogelsang and Petra Groß showed in a paper in *Nano Letters* that they could even create a free-standing electron gun by making use of plasmonic nanofocusing to induce electron emission. Ahmed was one of the referees of the paper and was very excited about this work. In his referee report, he asked Jan to directly measure the time resolution. Jan followed the advice and set up a new point-projection electron microscope. In proof-of-principle experiments, he managed to image photoemission from plasmonic nanoresonators in his microscope with a spatial resolution of 10 nanometers and a time resolution of only 20 femtoseconds. The results of these experiments will be published shortly and I very much hope that you, Ahmed, will like them.

Dear Ahmed, I wish to cordially thank you for all that you have done to support young scientists and science as a whole throughout the world. We will miss you.

Christoph Lienau, born in 1963, studied physics at the University of Göttingen, Germany. In 1992, he received a Ph.D. in physical chemistry for his research on the dynamics of elementary chemical reactions in molecular clusters. He then accepted a postdoctoral fellowship from the German Research Foundation and worked for two years as a research fellow in the Ahmed H. Zewail group at California Institute of Technology, Pasadena, USA, studying femtosecond dynamics of small molecules in solution. In 1995, he moved to Berlin, Germany, to become a member of the scientific staff of the newly founded Max-Born-Institute in the department led by Thomas Elsässer. Here, he initiated research on ultrafast nano-optics, combining femtosecond laser spectroscopy and nano-optical techniques to monitor and control the dynamics of ultrafast optical excitations in novel semiconducting and metallic nanostructures. In 2006, he became a full professor in physics at the University of Oldenburg, Germany. He is a Fellow of the Optical Society of America and, since 2015, chair of the semiconductor physics division of the German Physical Society. From 2009 to 2012, he served as the Director of Institute of Physics in Oldenburg. Since 2017, he served as the Dean of the Faculty of Mathematics and Natural Sciences in Oldenburg. His interests are in ultrafast and nano-optics with a particular interest in probing the motion of charges, spins, and nuclei in solid state and molecular nanostructures on ultrasmall length and ultrashort time scales.

18 Remembrance of Ahmed Zewail

Hua Guo*

In late 1992, I got a call out of the blue: "I am Ahmed Zewail from Caltech. I like your work on the photodissociation of methyl iodide and was wondering if you would like to visit me in Caltech for a few months." I was then a young second-year assistant professor at the University of Toledo, Ohio, and was stunned when I received his call. Ahmed was already one of the most respected scientists in the world and I could not believe that he was calling me and inviting me to visit Caltech. "I will be traveling too much in the next few months, including a trip to Israel for the Wolf Prize," he explained, "I would like someone to teach the Physical Chemistry class for me in my absence." I was simply thrilled to have the opportunity to teach in Caltech and do research with Ahmed.

My wife and I arrived in Caltech in early 1993 and stayed until the summer. My stay in the Femtoland was brief compared to others, but it had a huge impact in my career. At that time, Ahmed was in his prime — full of confidence and vision. His work ethic was legendary, but he topped it off with his infectious laughter and unique sense of humor. I had firsthand experience of how he ran his group, which was eye opening for me. We all knew then that Ahmed would get his Nobel Prize for his pioneering role in femtochemistry, but he

*Visiting Associate.
Email: hguo@unm.edu

continued to work hard on new and more difficult research projects, including ultrafast electron diffraction. I eventually published two papers with Ahmed, but it was the personal influence he had on me that has endured. He taught me, through his style of research, how to tackle difficult but solvable problems and to aim high and work hard!

Teaching in Caltech was not what I was prepared for. The students were highly motivated and knowledgeable, so I had to struggle to find things to talk about in class because they already knew everything in the undergraduate textbook. I shared these difficulties with Ahmed and he laughed out loud then said, "Hua, I don't know if you knew but in Caltech, undergraduate students are smarter than graduate students, who are smarter than postdocs. We professors are the dumbest!" He helped me to realign my teaching goals and adjust the pedagogy.

Twenty four years later, as I contemplate these unforgettable experiences with Ahmed, I began to realize how much influence he had on me. He was not just a brilliant scientist; he was also a kind and down-to-earth human being. Peter Dervan put it best — Ahmed made everybody feel that he was your best friend! We admire his scientific achievements but also miss his affections. He will always be with us, looking down from the pantheon of science with pride and happiness that his beloved scientific torch is carried on by so many of us.

Hua Guo obtained his B.S. and M.S. degrees in China. He studied theoretical chemistry with John Murrell, Fellow of the Royal Society, at Sussex University, UK and received his Ph.D. degree in 1988. After a postdoctoral appointment with George Schatz at Northwestern University, he started his independent career at the University of Toledo in 1990. By invitation of Ahmed Zewail, he spent a semester at Caltech in 1992 where he taught and collaborated with the Zewail group. He moved to the University of New Mexico in 1998 and is now Distinguished Professor of Chemistry. He was elected as Fellow of the American Physical Society in 2013. He currently serves as a senior editor for the *Journal of Physical Chemistry A/B/C*. His research interests include gas phase reaction dynamics and kinetics, photodissociation dynamics, surface reaction and heterogeneous catalysis, and enzymatic reaction mechanisms. He is the author of more than 400 peer-reviewed publications and has been cited more than 10,000 times.

19 Dear Professor Ahmed Zewail

Sang Kyu Kim*

Hi, Dr. Zewail.

Back in May 1993 when I was preparing for my Ph.D. thesis at Berkeley, you called me from Caltech asking if I can join your group as a postdoctoral researcher. Kindly, you offered me a very honorable Arthur Amos Noyes postdoctoral fellowship. When I heard your voice on the phone, I am now telling you, I was so much thrilled and excited about joining your laboratory. On the other hand, I did not give an answer right away at that time. Frankly, I was somewhat hesitant to go to Caltech for many reasons, one of which was that I was a little bit scared to start new things. However, joining your laboratory turned out to be one of the best decisions that I have made so far in my careers. I thank you so much for giving me the opportunity.

Now, I am proud and very much honored that I was able to be one of the many coherently excited members in your great personnel wavepacket. Through the exploration of excited-state proton-transfer and/or unimolecular reaction dynamics, I was involved in several very interesting chemical subjects with truly outstanding colleagues. I was lucky enough that I could be part of beautiful works done with Dr. Soren Pederson, Dr. Jun-Kai Wang, Dr. A. Douhal, and Dr. Baskin.

---

* Postdoctoral Research Fellow.
Email: sangkyukim@kaist.ac.kr

Our coherence as collaborators was quite constructive in many ways. I should tell you that I was always inspired by your dauntless challenges out of perseverant and consistent curiosities in sciences.

Dr. Zewail, you have done a lot in basic chemical sciences. You have made it possible to look at chemical reactions in real time by all means so that we can get deep insights into complex nuclear motions occurring on quite complicated potential energy surfaces during chemical reaction processes. Your adventurous wavepacket has never been stationary. Rather, it has explored many different phase spaces of biology, physics, and material sciences. And I am very much sure that your wavepacket will never disappear in the future. Rather, it is being consistently created and evolved in a number of prestigious laboratories all over the world.

The lifetime of a human being is as short as a femtosecond. And yet, you have shown us how such a short life can be so fruitful and helpful to mankind. Nowadays, everyone is connected in real time and I hope that our Zewail family can be coherently excited anytime and anywhere because of our mentor's great spirit.

With all of my respect and love,

Sang Kyu Kim (postdoctoral researcher from November 1993 to February 1996)

Professor of Chemistry, Dean of Colleges of Natural Sciences at KAIST (present)

Seoul National University (B.S., M.S.), University of California at Berkeley (Ph.D.), Caltech (postdoctoral research), Inha Univeristy (1996 to 2003), KAIST (2003 to present).

20 My Memories of a Giant

Dongping Zhong*

Ahmed Zewail has left us for almost a full year. Throughout the year on various occasions, I have been in contact with many of Zewail's former group members and one common theme is that we have seen him with a big smile in our dreams. He will be remembered dearly as he has had such a huge impact on so many people's lives. The recent request by the group members to put together a commemorative book to collect cherished memories of our personal interactions with Ahmed, one of the most influential scientists in the past 50 years, is both touching and visionary. This book will be historically important as it not only documents the snapshots of his major scientific breakthroughs at Caltech, but also reveals many personal encounters and stories from first-hand experiences. These stories revolve around interactions with the scientific giant that is Ahmed, and cover 40 years at Caltech from 1976 to 2016. I recently wrote a contribution, My Time with a Giant, in another memorial book[1] and shared my past experiences from the 24 years that I had spent in the United States working and interacting with Ahmed. Here, I share several more memories of him during my scientific career and daily life.

---

* Graduate Research Assistant/Postdoc/Visiting Associate.
Email: dongping@physics.osu.edu

[1] *Personal and Scientific Reminiscences-Tributes to Ahmed Zewail.* Majed Chergui, Rudolph A Marcus, John Meurig Thomas and Dongping Zhong, Eds., World Scientific, London, 2017.

I went to Caltech in December 1993 and immediately joined the Zewail group. After working with Dr. James Cheng for several months, we were able to successfully detect an iodine atom in real time and performed our first experiment on the dissociation of iodobenzene (C_6H_5-I). The aim of this experiment was to observe two types of C-I dissociation — direct and indirect. The direct one comes from the dissociative excitation of the C-I bond ($n\sigma^*$) and the indirect one results from the intramolecular vibration redistribution of degenerately excited C_6H_5 moiety ($\pi\pi^*$). We monitored the product of iodine atoms with different flying speeds and clearly separated the two reaction channels with distinct time scales. We got very excited and planned to do a series of chemical substitutes on the benzene ring to observe different indirect C-I dissociation times and thus understand energy redistribution rates between two different modes. When we went to report these outcomes to Ahmed, what he said surprised me at first. I still remember to this day what Ahmed said: "Dongping, this can lead to your Ph.D. thesis but I am not interested. Leave this to other people and let's move on to other systems." It was only some time later that I fully understood why he said that. From iodobenzene, we had obtained the actual time scales of the two coupling modes and thus understood the framework of these types of reactions. All other similar studies could only be additively incremental and would not lead to more knowledge breakthroughs. Instead, we moved on to examine a series of new reactions: the three-body dissociation of I-Hg-I, the sequential dissociation of I-C_2F_4-I, the nucleophilic substitution reaction of C_6H_5-I···Cl_2, the four-centre reaction of CH_3-I···I-CH_3, and the famous charge-transfer bimolecular reaction of benzene C_6H_6···I-I. The studies of those reactions were rich, new concepts were introduced, and new knowledge was obtained. We published 11 papers on various types of elementary reactions with atomic details. Clearly, Ahmed was right. He was truly a visionary — he pioneered Femtochemistry and quickly expanded the concept and methodology into all branches of chemistry. As stated in the Nobel Prize press release in 1999, "Professor Zewail's contributions have brought about a revolution in chemistry and adjacent sciences."

After I graduated in 1999, Ahmed asked me to stay at Caltech as a postdoctoral researcher. He gave me complete freedom to explore biological dynamics. I was interested in electron and energy transfer in proteins and we first studied energy transfers between a natural amino-acid tryptophan with an iron-sulfur cluster in protein rubredoxin. We unintentionally observed water motions around the protein from tryptophan fluorescence emissions. Again, Ahmed's insight was visionary and he asked us to focus on the hydration dynamics of proteins, a topic central to protein science. Later, I focused on this in my independent research extensively using protein engineering. Even until today, protein hydration dynamics remains a hot topic and is still not fully understood at the molecular level. After I moved to Ohio State University in 2002, Ahmed and I continued to collaborate on this topic and we published three more papers. I visited him several times (Photo 20.1) and we always had long rewarding discussions spanning from the morning to evening. Ahmed's brilliant vision, insights, and clarity on the complexity of hydration dynamics have reshaped the way we think about the field of protein hydration.

After 2002, we always met at the biennial Femtochemistry meeting (Photo 20.2). Back in those days, we often discussed about the frontiers of the ultrafast field as well as various world affairs. He was amazed at China's rapid development and asked me how China did it and what really happened there. Every time, he was not satisfied with what I told him. Even to this day, I wish I could have given him a satisfactory answer. I am sure these questions were driven by his concerns about Egypt. If China can do it, why not Egypt?! In the summer of 2004, we, with his wife Dema, visited China for about two weeks. I arranged for him to visit three universities in three quintessential Chinese cities — Peking University in Beijing (Photo 20.3), Huazhong University of Science and Technology in Wuhan, and Fudan University in Shanghai — to get a general feel of the recent changes in China. Ahmed was clearly excited and he was eager to see and understand everything from the daily life of the Chinese people to the upper levels of Chinese science policy. We visited many places in Beijing (Photo 20.4) and also met with the leaders of

Photo 20.1. Taken with Professor Ahmed Zewail when Dr. Zhong visited Caltech again after 2002.

the Chinese academy (Photo 20.5). He received an honorary doctoral degree from Peking University; a prestigious honor that is usually given only to foreign country leaders or famous statesmen. He had signing activities in the three universities for his translated book premiere, *Voyage Through Time*. He was so popular; thousands of people attended his talks and hundreds of students stood in line for his signature. His book has certainly become one of the most influential scientific biographies in recent years. He told me that he liked China and promised to visit again. From the beginning, he had been supportive of the development of ultrafast science in China and also promoted Beijing to host the 9th Femtochemistry Meeting in 2009. Two arrangements in 2009 and 2016 were made for him to visit China again, but he could not make it due to his national duties within the Obama Administration for the former and his illness for

Photo 20.2. Taken with Professor Ahmed Zewail in front of the president's house of Magdalen College in Oxford after the 8th Femtochemistry Meeting in 2007.

the latter. The Chinese people remember him dearly. I am sure he would be very happy if he knew that the 14th Femtochemistry Meeting will be in China again in 2019 and, this time, it will be held in his favorite city, Shanghai.

Another memorable and touching event was in 2008 after I was tenured at Ohio State University. I invited Ahmed to visit our university and give two named lectures, the Robert Ross Lecture in Biophysics and the Alpheus Smith Lecture in Physics. I was honored to host him, especially because I was holding the Robert Smith Professorship which bears the name of Alpheus' son. He arrived in Columbus around 6.00 p.m. after a long flight and wanted to see my family even though he looked very tired. He brought my two boys

Photo 20.3. Taken with Professor Ahmed Zewail in front of an ancient house in Peking University where Ahmed and Dr. Zhong stayed during the visit.

his signed Nobel Prize pictures and they were so thrilled to listen to Ahmed's edification (Photo 20.6). We had a wonderful Chinese dinner at home with his favorite dish, Peking duck. He then gave a beautiful public lecture titled "Mysteries of Time" (Photo 20.7), covering a wide period from ancient Egypt to our modern era. My colleagues later told me that was the best lecture they had ever heard in recent years. In the four days at the university, he was extremely busy talking to many faculty members and students. We had a wonderful time and even managed to enjoy some cigars together at night.

In the two years of 2010 and 2014, I went to Caltech and spent a significant amount of time on femtosecond-resolved 4D electron microscopy, which was the recent major breakthrough in the Zewail group. For the first time, people were able to image complex structural changes in space and time simultaneously with atomic resolution. He completely opened another new field using ultrafast

Photo 20.4. Professor Ahmed Zewail and his wife Dema Faham during a visit to Beijing in 2004.

Photo 20.5. Professor Ahmed Zewail and Dr. Zhong having lunch with President Yongxiang Lu (third right) of the Chinese National Academy of Sciences in June 2004, accompanied also by Professors Jie Zhang (second left), Jinghua Cao (far left), and Qihuang Gong (far right).

Photo 20.6. Professor Ahmed Zewail at Dr. Zhong's house with his two boys during a visit to Ohio State University in 2008.

electrons and almost made another voyage to Stockholm! Ahmed was a visionary leader, a brilliant scholar, a giant scientist, and a deep thinker. His impact on science was profound and far-reaching, and his love and dedication to his motherland through the use of his own life to advance science and technology has made him a truly faithful son of Egypt. People will certainly remember him forever. Even today, as I sit in my office and pen these memories, he is still right in front of me and always in my heart.

Photo 20.7. The public Smith lecture given by Professor Ahmed Zewail at Ohio State University in 2008.

Dongping Zhong received his B.S. in laser physics from the Huazhong University of Science and Technology in China and his Ph.D. in chemical physics under Ahmed H. Zewail from the California Institute of Technology in 1999. For his Ph.D. work, Dr. Zhong received the Herbert Newby McCoy Award and the Milton and Francis Clauser Doctoral Prize from Caltech. He continued his postdoctoral research in the same group at Caltech with a focus on protein dynamics. In 2002, he joined Ohio State University as an assistant professor and is currently Robert Smith Professor of Physics and Professor of Chemistry and Biochemistry. He is a Packard Fellow, Sloan Fellow, Camille Dreyfus Teacher-Scholar, Guggenheim Fellow, American Physical Society Fellow, American Association for the Advancement of Science Fellow, as well as the recipient of the National Science Foundation's CAREER Award and the Organization of Chinese Physicists and Astronomers' Outstanding Young Researcher Award. His research interests include biomolecular interactions and dynamics using ultrafast photon and electron methods.

21 Recollections of My Days with Professor Ahmed H. Zewail

Po-Yuan James Cheng*

I do not remember the exact date, but it was probably somewhere between 1988 and 1989 when I saw Professor Zewail for the first time. He was speaking at an American Chemical Society National Meeting. I was a young graduate student back then, and I think I joined the crowd attending his plenary talk. The large auditorium was full of people and excitement, and I was probably either standing in the aisle or sitting in the back, listening to his lecture from a distance. The man on the stage was talking passionately with an exotic and charming accent, and the talk was most likely about some initial results of one of the very first femtosecond experiments conducted by his group. Although I could not fully understand the contents of his talk at the time, the topic and the way he delivered the science attracted my attention immediately. I would never have imagined at that moment that the experiments he was presenting would eventually earn for him the Nobel Prize, nor would I have imagined that I would have the privilege to work with and learn from this great scientist in the future.

Several years later in February 1994, I found myself sitting in his spacious office at Pasadena for the first time after I joined his group as a postdoctoral research associate. I was a little nervous, of course, after being "cautioned" by quite a few people. Nevertheless, to my

* Postdoctoral Research Fellow/Visiting Associate.
Email: pycheng@mx.nthu.edu.tw

full surprise, this man was indeed quite affable and even a bit humorous. He first greeted me with a very warm welcome, enquiring about how I had arrived and if my family was settled. Then, he started to talk enthusiastically with all his smiles about the experiments he had in mind for me to carry out in Femtoland Room 048. It must have been the way he talked that enabled me to feel his contagious passion for science immediately. I was all motivated but also a bit anxious; although I had a lot of experience with nanosecond laser spectroscopy and mass spectrometry before I joined his group, I did not have any hands-on experience with the ultrafast laser back then. Many years later when I thought about this, I was always puzzled as to why he had so much confidence in me being able to handle the experiments he wanted me to carry out.

Soon after I arrived, Dong-Ping, who was a young graduate student back then, joined me to work in Room 048. AZ (what we called him in the basement of the Noyes Laboratory) gave us some general directions and allowed us to delve into the experiments on our own. I recall that our progress was a bit slow in the beginning, but he patiently waited, making only occasional enquiries for any new results. After several months of hard work, we managed to acquire some preliminary data to show to him in his office. We were a little nervous because we were not sure if the data was exciting enough for him. It turned out that Dong-Ping and I thoroughly enjoyed that first discussion, not only because AZ generously complimented us, but also because he was even more excited about the results than we were! Occasionally, AZ would come unannounced to our laboratory to watch us running experiments and talk to us. This especially happened whenever we told him that a new experiment was working and we were getting some nice data. He enjoyed watching data emerge from the machines in front of him when everything was running in the laboratory. He told us that he never quite believed the data unless he could see them coming out from the instrument with his own eyes. That is a rule that I have continued to follow to this day.

Very often when the experiments were going smoothly at a seemingly unstoppable pace, Dong-Ping and I could not resist the temptation to continue working past midnight. A few times when we were working that late, AZ would call us at one or two o'clock in the morning with a sleepy voice to ask how the experiment was going. This may sound unbelievable to some people whom I talked to years later because they thought AZ was probably calling to check on our dedication or work ethic. But, the truth was that he knew we were working late at night, so he thought he should call to encourage us and to let us know that he cared about how we were, even though he had to wake up after midnight. Now, I am the age he was then, and I wonder if I would bother to wake up at one o'clock in the morning to make a call if I knew my students were working at midnight!

When we were having some bottlenecks in our experiments, he would sometimes call us to meet upstairs at his office and talk about the troubles we were facing. In those situations, the air was sometimes a little more tense than usual, but he never really was upset with us. Instead, he would patiently listen to our excuses, with his signature bewildered smile, and then offer some suggestions, always encouraging us to never give up hope. Many years later, after he had recovered from surgery and returned to work in 2014, I sent him an email to encourage him with the same words.

Occasionally, when he was not too overloaded with his work, he would invite me for lunch at some of his favorite restaurants on Lake Avenue. During those lunches, we also often discussed the ongoing experiments in the laboratory or made some outlines of the papers we were about to write. After that, AZ would sometimes also talk about things other than science. In particular, he was quite interested in political and economic developments in Taiwan where I am from. Back in those days between 1995 and 1996, Taiwan was undergoing a democratic transformation for the better. He enquired, on many occasions, about the political situation in Taiwan and tried to make comparisons with his own motherland. I could tell that he was very eager to do whatever was needed for Egypt. I was not surprised when I heard that he rushed back to Egypt during the 2011 revolution in an attempt to help the country get on the right track. He replied to my email in 2014 after his surgery and said, "It is the efforts in science and the global affairs that help me to conquer barriers, and of course with the essential support of my family." I truly believe that his efforts will not be in vain; the seeds he planted in the soil of Egypt will one day bring his beloved motherland to a renaissance.

I stayed in Zewail's group for only about two and a half years. When I recall that period many years later, I realize that they have been the most splendid and rewarding moments of my life. Although working in his group was demanding, it was also full of excitement and passion for making new discoveries; under his mentoring, everyday was a challenging scientific exploration. I am tremendously grateful to have had the privilege to learn from a great scientist like him. He not only showed me the beauty of science but also exemplified how a scientist makes great contributions to the world.

Po-Yuan James Cheng received his Ph.D. in 1990 at the University of Georgia under the supervision of Michael A. Duncan. He did his first postdoctoral work at the University of Pennsylvania with Hai-Lung Dai. In February 1994, he joined the Zewail group as a postdoctoral research associate and worked until July 1996 before he returned to Taiwan. He is now a professor of chemistry at the National Tsing Hua University in Taiwan.

22 A Secretary's Recollections

Mary Sexton*

I worked for Dr. Zewail for eight years as his assistant and his secretary. One of the things that used to amuse me about Dr. Zewail was his amazing memory and apparent ability to remember every single piece of correspondence he had ever received or sent and when. He would often say to me that he wrote (or received) a letter to (or from) "so and so" in 1987 (or whatever date) and ask me to get it from the files. Of course, it was always there. He did this repeatedly over the years. I frequently told him that I thought he must have a photographic memory and he would always look at me with a twinkle in his eyes and a little mischievous grin and never answer me. I simply could never understand how he was able to do that.

Although it could be a challenge to work for such a busy, focused, and brilliant man, I thoroughly enjoyed my job. Just being around him and seeing all his accomplishments and how he handled his work was fascinating. I also enjoyed the members of his research group who were always so pleasant and kind to me and it was impressive to be in the midst of so many talented young people with all their ambitions.

Of course, it was an amazing experience to be there when Dr. Zewail won the Nobel Prize. It was certainly one of the most

---

* Secretary.
Email: marycsexton@outlook.com

Photo 22.1. From left, Mary Sexton, Karen Hurst, Dr. Zewail, Jeanne Rademacher, and Marjorie Miller; four of his administrative/clerical workers on the day of the Nobel Prize announcement in the garden at the Caltech President's home.

chaotic work scenarios I have ever experienced! On the first day alone, hundreds of phone calls, emails, and faxes poured in which was incredible, and later all the letters, strange requests, reporters, and so on followed which was also unimaginable. Finally, we had to go through all the preparations for the ceremony. It was an unbelievable experience and I felt privileged to be a small part of such an event and to witness history in the making.

He was extremely organized. I also tend to be that way in my work, so I always found it amusing to look at his desk and observe how very organized it was and then look at mine which was also normally the same, and then try to decide who was the most organized. His birthday was February 26 and mine was February 28, so we were born under the same astrological sign, but this was the only similarity I could find between us — clearly there were no others!

Another side of Dr. Zewail which I admired and respected was his humanitarian interests and goals. In particular, although America was his adopted country, he still remained loyal and dedicated to Egypt, his native land, and worked hard to improve conditions for the people of Egypt. He did not forget his roots.

It was also quite interesting to help with the work on his autobiography. Most of his administrative staff helped with the typing and editing. It was fascinating to learn all about his early life and the circumstances that eventually made him the man he became.

I would like to end with my thoughts of a side of him that many, but possibly not all, would have been aware of. During my time with him, I went through two very difficult periods — one when I had cancer and one when my mother was critically ill (she subsequently passed away) — both of which severely affected my ability to carry out my duties as his secretary. However, he had a very compassionate side and could not have been more understanding, always gladly agreeing to anything I needed and kindly offering his help. Working for him could be fun, challenging, and fascinating, but the thing I will remember most is his compassion for people.

There were two ladies who were a huge part of Dr. Zewail's administrative staff and unfortunately, both have passed away. I strongly feel that they should be a part of this book and, in their absence, I would like to add some thoughts on their behalf.

Mrs. Tina Wood worked for Dr. Zewail as his secretary for many years, beginning from the time he first came to Caltech. She, thankfully, was able to attend the marvelous tribute to him on the occasion of his 40 years at Caltech which she thoroughly enjoyed. I had the pleasure of visiting with her on that occasion as well as many times in the office when she joined us for lunch or other occasions, and also conversing with her often on the telephone. Unfortunately, Tina is unable to make what would have been an important contribution to this book as she passed away in 2016. It is a shame, because she would have had so many interesting things to add and would have enjoyed this sharing process immensely. In her absence, I would like to acknowledge how deeply she admired Dr. Zewail and followed all his work and accomplishments long after her retirement. She often entertained me with stories of working with him, and since she was with him from the beginning, I know that she was a tremendous help

Photo 22.2. From left, Mary Sexton, Tina Wood, Dr. Dema Zewail, and Dr. Zewail.

Photo 22.3. Shortly after the Nobel Prize announcement, tee shirts were made to commemorate the occasion. Pictured are Dr. Zewail, his research group, his administrative/clerical group including Sylvia Jacoby (second from left, first row), and various friends and co-workers from Caltech.

to him in learning the ways of Caltech and, in some cases, the cultural ways of the United States. I know he respected her greatly and always had her join the rest of us on many occasions.

Ms. Sylvia Jacoby worked many years for Dr. Zewail as his assistant in his capacity as editor of the journal, *Chemical Physics Letters*. Her efforts in that regard were monumental — running that office, keeping track of all the manuscripts which were submitted for review, and following through all the many steps to be accomplished before they were ready for publication. She was a hardworking and dedicated employee and person whom Dr. Zewail highly respected and I know, as a friend of hers, the high regard she also held for him. She was a great fan of the television program *Jeopardy* and I have never forgotten how excited she was when Dr. Zewail became one of the questions/answers on that show! After he won the Nobel Prize, more clerical work was needed and she joined the rest of us in our effort to keep things organized. She, also, would have had many interesting recollections to add to this book.

23 Hoy and Ah

Hyotcherl Ihee*

As I write this piece, I realize that my memories about AZ do not form a single coherent story but are rather a collection of smaller fragments of memories, probably due to the elapsed time and my evaporating memory. Here, I aim not to put together a coherent narrative, but rather to present those fragments as an authentic mosaic of my recollections.

I clearly remember his voice — deep, low, and often full of passion. Sometimes his voice was intimidating, but somehow I liked it when he called my name. My first name, Hyotcherl, could not be easily pronounced by non-Koreans. AZ was no exception. Instead of "Hyotcherl," he always switched the "y" with "o" and pronounced it as "Hoy-chul." Not long after I joined his group, he called on Chuck Williamson, the senior student working on the ultrafast electron diffraction project, and suggested the idea of calling me "Hoy" instead of his version of the full first name "Hoy-chul." Chuck then replied to him, "Well, that sounds like a good idea, but I am afraid that people may call you 'Ah' instead of 'Ahmed.'" AZ quickly shelved this idea and Chuck saved my name.

I first met him when I had just arrived at Caltech to start my Ph.D. program and I had been trying to find a Ph.D. mentor. At that time, my English was even worse than today. I had to go through some

* Graduate Research Assistant.
Email: hyotcherl.ihee@kaist.ac.kr

trouble when I visited chemistry department to find out the location of his office. No matter how many times I said, "I would like to meet Professor Zewail and know his office location," the department secretary could not understand me. So I wrote down his name and then she said, "Oh, you mean Professor Zewail." Later, I found out that my pronunciation of "Zewail" sounded more like "Jewail" to her because my "Z" pronunciation was not good. Korean phonetics do not include "Z" sounds and the only sound that comes close is the "J" sound. At that time, I also had hearing problems and I often had to ask people to repeat themselves. So it was not a surprise that I could not really understand what AZ was saying when I met him for the first time in his office. He suggested going with him to a Starbucks cafe and we talked while walking there. Although I could not understand him fully, I could feel that he was full of energy and passion. Having such a high level of passion is one of the two most important things that I learned from him and continues to guide me along my scientific career. Anyway, I guess he could not understand much of what I was saying in broken "Konglish" either!

Indeed, this turned out to be the case. Fast forward to the day I successfully defended my Ph.D. — the AZ research family had gathered, as per tradition, to celebrate with champagne. The person who had just passed the defense was to write the date and his or her signature on the bottle of the champagne and AZ kept the bottles in his office. As I received a good round of felicitations from everyone gathered, AZ said, "Hoy-chul did very well. You know I have one thing to confess. When he came to meet me for the first time, actually I could not really understand what he was saying. But I could see a strong passion in him and that was why I accepted him to my lab." People enjoyed his remark and laughed. Then I replied, "Actually I also have to confess. On the first day, I could not understand what you were saying, either." My reply drew just as much laughter!

If we think about that first day, a professor from Egypt and a student from Korea were talking to each other without really understanding what the other was saying. But AZ did not look at me only at face value and instead tried to see the potential in me. I really appreciate it and will always remember this. He took a risk and made a bet on a student from Korea who was struggling to even make himself understood. I would like to believe that his bet was successful as I was able to accomplish some of his research goals on ultrafast electron diffraction in the gas phase.

I mentioned that there are two important things that I learned from AZ which are still helping and guiding me in my scientific career. One is passion, his enthusiasm for science. The other is the audacity to challenge common sense and traditionally held ideas as well as his persistence to surpass conventionally accepted concepts and seek novelty. Throughout my study under his supervision, I learned that he was never discouraged by what people were saying against some of his ideas. When we wrote up a manuscript, he liked to sit down with his students and postdoctoral researchers and correct the manuscript line-by-line with a red pen. He really enjoyed it. Sometimes, we tried to explain that our observations were somehow against some of the previously published data or proposals. Instead of trying to make a compromise between our work and the literature, he almost always tried to emphasize the novelty of our findings from a new perspective or concept. Later, I adopted his audacity of being unafraid to challenge the status quo and it helped me greatly when I started my own independent career as an assistant professor at Korea Advanced Institute of Science and Technology. At that time, although an early version of time-resolved X-ray scattering on liquid solution had been done, people did not really believe that highly detailed structural information approaching atomic resolution can be obtained from such methods. Even the well-known textbooks on X-ray diffraction made this point clear. However, due to the audacity I learned from AZ, I mustered the courage to be undeterred by such a common and traditional idea and did my own time-resolved X-ray solution scattering experiments on complex molecules to extract detailed structural information on molecular reactions from the time-resolved X-ray solution scattering data. It worked! People failed to try not because it was particularly difficult but simply because, based on conventional wisdom, they thought it would not work. Also, because I had to suffer so much from trying to observe the small difference signal of ultrafast electron diffraction in the gas phase, the difference signal from ultrafast X-ray scattering in the liquid solution phase looked relatively much larger (although the analysis *does* get complicated due to the presence of solvent molecules which are absent in the gas phase). So, my success with time-resolved X-ray solution scattering (also known as liquidography) is owed to AZ, although he was not directly involved in it.

The rather sudden passing of AZ last year was a sad reality that I could not accept easily and I still hate to accept. It made me numb for a while. I can still hear his voice calling his version of my name, Hoy-chul. Although he is not here physically any more, his legacy to me — passion and audacity — will continue to guide me through my scientific journey.

Hyotcherl Ihee obtained a B.S. from Korea Advanced Institute of Science and Technology (KAIST) and a Ph.D. from Caltech in 2001 under the supervision of Professor Ahmed Zewail and working on the development of ultrafast electron diffraction. He then joined Professor Keith Moffat at the University of Chicago as a postdoctoral fellow working on protein structural dynamics using time-resolved X-ray crystallography. In 2003, he became a faculty member at KAIST, where he has made significant contributions to the establishment of time-resolved X-ray liquidography, which is a state-of-the-art method for visualizing 3D molecular structures of reaction intermediates in the liquid and solution phases. Using the technique, he has elucidated the structural dynamics and the mechanisms of various haloalkanes, photocatalysts, and proteins. His research is focused on understanding the structural dynamics of a wide diversity of molecules ranging from small molecules to proteins using time-resolved X-ray liquidography, time-resolved X-ray crystallography, and time-resolved optical spectroscopy. Professor Ihee proposed an idea to capture the 3D structure of rarely populated states such as the transition state by using single-object scattering sampling. In addition, he has revealed the relationships of catalyst-substrate association for diverse metathesis Mo and Ru catalysts using time-dependent fluorescence quenching spectroscopy.

24 Time Working with AZ

Chaozhi Wan*

Femtoland is the name of our research group in Caltech and Professor Zewail was the leader. We simply called him "AZ" most of the time and his research group and laboratories followed suit being known as the "AZ group" and "AZ labs." The time I spent working with AZ was the longest period of my scientific career.

AZ was a good mentor and friend to me. When meeting with him, I always called him Dr. Zewail or Professor Zewail, since this is the way to show respect according to Chinese traditions that the family name instead of the first name should be used in most cases. AZ often complained that many people could not pronounce his name right, and I believe I am one of these people! There is a funny story about his name. During a group meeting after AZ visited China, he happily mentioned that his book *Voyage through Time* had been published in Chinese and he asked me if I knew his name in Chinese. I told him that I knew at least two versions of his name in Chinese. He got really confused and asked me to write his Chinese name, so I wrote two Chinese names on the blackboard of the meeting room. Then the group members from China and Taiwan, like Dongping and Ruan, added more versions of his name in Chinese. From these five to six names, we couldn't agree on the best Chinese name for him. Finally, AZ had to give up this topic.

---

* Senior Scientist.
Email: wan@minioptic.com

The sad news last August from Spencer in Caltech deeply shocked me. The air felt frozen and sadness filled my heart. We have lost Ahmed Zewail — the great scientist, Nobel laureate, mentor, and friend — forever. I could not believe the message as I thought AZ had recovered well from cancer and I wanted to visit him soon. It was July about three weeks earlier when I had just met with Spencer in Caltech to discuss the pulse width measurement using one of my autocorrelators and learned that AZ seemed in good shape after undergoing new treatment. In fact, it felt like business as usual because AZ's secretary, De Ann Lewis, told me in our telephone conversation that he even asked for more than one year's warranty time when ordering the MiniScan-200PS autocorrelator. I thought AZ would continue doing his scientific research for many years to come.

I have worked in the AZ research group at Caltech from 1994 to 2008. This is a long period of time and I feel lucky and privileged to be able to work with a giant scientist. My first time knowing AZ was in 1994 when he visited the University of Kansas (KU) and gave a scientific seminar in the chemistry department. He was invited by Bob Bowman, a KU assistant professor at that time and a member of the AZ group. At that time, I was doing research in Professor Carey Johnson's group in KU's chemistry department. Bowman's laboratories and ours were on the same floor in the same building. During that time, I had worked to build a Ti:Sapphire femtosecond laser in our laboratory which could generate a <30 fs pulse with good stability and easy operation. Femtosecond Ti:Sapphire laser, a solid state laser, is much better than a femtosecond dye laser. AZ visited our laser laboratory with great interest. We had some discussions about Ti:Sapphire lasers and AZ asked questions about the laser pulse width and the wavelength bandwidth. The performance of the Ti:Sapphire laser must have given AZ a good impression. After AZ finished his seminar talk, I approached him and asked if he had a position open so that I could join his research group in Caltech. To my surprise, AZ immediately answered, "Yes, I have a position open. You could join us." Then he asked me to send my resume to him and asked me to contact his secretary Mary to start the job application process.

After I came to Caltech, I was really surprised to learn that all the femtosecond lasers in AZ's laboratory were dye lasers. These were dye colliding-pulse oscillators which generated femtosecond pulses that were next sent through several amplification stages to increase the femtosecond pulse's energy for various experimental purposes. Dr. Zewail had made great scientific discoveries and gained his fame using these femtosecond dye lasers. However, these dye lasers have stability problems caused by the bleaching of dye molecules and thermal fluctuation. When using a femtosecond dye laser, experimenters have to do a lot of alignment adjustments and change the dye each day, thus limiting the time available to actually do experiments. Compared to the femtosecond dye laser, the femtosecond Ti:Sapphire laser is superior in stability and easier to operate. In fact, the femtosecond Ti:Sapphire laser is almost a fully hands-free, turnkey system. I soon realised that part of my work would be to upgrade the femtosecond dye lasers to femtosecond Ti:Sapphire lasers in the research laboratory. My first research project in AZ's laboratory was to study the vibration dynamics of iodine molecules under high pressures of up to 3000 bar. For this research project, we still used a femtosecond dye laser since it took some time to order a femtosecond Ti:Sapphire laser and wait for it to arrive. Our first Ti:Sapphire amplifier laser was ordered from Clark with a short pulse width to about 50 fs while Spectra-Physics' Ti:Sapphire amplifier laser had a > 100 fs pulse width at that time. Soon, we had another two femtosecond Ti:Sapphire amplifiers for laboratory 051 doing ultrafast electron diffraction and laboratory 05A for the study of biomolecule dynamics.

One of the experimental research projects using the new femtosecond Ti:Sapphire laser in laboratory 05A was to study the charge transfer dynamics of DNA double strands. There were arguments about the charge transfer rate of DNA at that time. One group thought the DNA behaved like a conductor so the charge could flow in DNA very fast, while another group believed DNA to be an insulator so the charge transfer could not flow in DNA. AZ chose this project of charge transfer in DNA because he believed we could solve this problem by measuring the real time rate of charge transfer in DNA double strands using our new femtosecond Ti:Sapphire laser system. Before this, people had studied the charge transfer rate by steady state methods but not by measuring the rate in real time. Through real-time measurement of the rate, our study would be key to understanding the behaviour of charge transfer in DNA. AZ chose his research projects by two criteria: we can either do it best or do it first. For this project on the charge transfer in DNA, we collaborated with Professor Jackie Barton's group in the Caltech Chemistry Department. Jackie's group had previously done quite a fair bit of work on the charge transfer in DNA using steady state measurements. They made very well-prepared DNA samples with the charge donor and acceptor locked in the DNA double strand. The charge donor and acceptor are separated by the spacer base pairs which do not generate or accept charges. The distance between the donor and acceptor is known depending on the number of the spacer base pairs from 0 to 4. We used two femtosecond laser pulses in the experiments which were generated by two optical parametric amplifiers pumped by the femtosecond Ti:Sapphire amplifier. We used an excitation laser pulse to generate the charge from the charge donor and used a probe laser pulse to measure the spectral change of the charge acceptor as a function of time between the excitation pulse and the probe pulse. From the real-time dependent transient absorption spectra, we obtained the charge transfer rate in the DNA. The experiments were time consuming due to the signal being very weak and sometimes we had to continue a measurement for up to a week just to get one trace.

Laboratory 05A mainly concentrated on the ultrafast dynamics of biomolecules in water solutions, which was different from most of AZ's other laboratories that studied small molecules in the gas phase or electron diffraction in high vacuum. The 05A sample systems included DNA, RNA, and proteins. We used transient absorption and fluorescence up-conversion spectroscopy methods to study the ultrafast dynamics in the biomolecules. In these pump-probe experiments, we used a pump laser to excite molecules and then used another laser with a time delay to probe the spectrum change induced by the pump beam. Our femtosecond laser system allowed us to study ultrafast dynamics from a 100 fs time scale to relatively slower processes of about 10 nanoseconds. One project in our study was related to protein folding and unfolding, which is important for understanding protein function. Many studies in theoretical calculations and experimental investigations had been done and temperature jump had been identified as a useful method to investigate protein folding and unfolding. In temperature jump experiments, a pump laser quickly raises the temperature from 10°C to 20°C and then a probe laser is used to monitor the relaxation process of protein to the equilibrium temperature. However, most previous temperature jump experiments were done on relatively longer time scales from microseconds to milliseconds or longer. AZ was far more interested in the initial kick off of the protein folding and unfolding process. Therefore, we performed temperature jump experiments with femtosecond time resolution to investigate the early processes of protein relaxation. The temperature jump experiments reveal that the initial temperature rise and the following water relaxation finished within a 20 ps timescale and the initial protein relaxation takes as long as a few nanoseconds.

AZ really loved science. His enthusiasm and passion for science set a high standard for others. He worked most of Saturday and there were stories that he could call in midnight to discuss experimental results or he might have pushed people very hard in his group. I like to use some of my experiences with AZ to show another side of him. The first thing I would like to mention is that I didn't work on Saturday even though AZ himself and most group members did. After entering middle school, my son Jack often had school activities on weekends which kept me very busy. Therefore, I talked to AZ and told him that I wouldn't be able to come to the laboratory on weekends. AZ understood my needs and didn't say anything except that he might need me sometimes on weekends and that he would inform me in advance in these cases. I believe I am one of the very few people in the AZ group who didn't work on Saturday. The issue here though isn't so much about whether people were obligated to work on the weekend, but rather that we belonged to a work environment where every person worked hard for science to the best of their ability, including AZ himself.

Another story is about Aiguo, a former postdoctoral researcher who is now a professor in China. Within a short time after Aiguo joined this group, AZ appeared to be getting increasingly angry with Aiguo as his experience was not what AZ expected as recommended by Aiguo's former professor in Europe. The situation got so bad for Aiguo and he feared that he would be forced to leave Caltech early. It was even worse that, due to language barriers, Aiguo could not explain things clearly himself. However, AZ did not jump to conclusions and sought my opinion about Aiguo's work. I gave AZ my opinion that Aiguo might work well in laboratory 05A with me since he should be especially competent in dealing with biomolecules. After our talk, AZ went along with my suggestion and got Aiguo to join me in laboratory 05A and continue his work, especially since the whole situation was not really Aiguo's fault and AZ wanted to help rather

Photo 24.1. The author (third from left) and AZ on the beach in August 2005 at the group retreat at El Capitan Canyon, near Santa Barbara with other group members (from left) Hern (Daniel) Paik, Ramesh Srinivasan, De Ann Lewis, and Spencer Baskin.

than damage a young person's scientific aspirations during the early stages of his career. The result is that Aiguo successfully finished his stint in Caltech and had a paper published with AZ. The experience in Caltech, I believe, should greatly help Aiguo's career in science later. I truly believe AZ is not only a great scientist but also has a kind heart to help other people.

Chaozhi Wan owns and works at Minioptic Technology, Inc. which is located in California since 2008. He was a research member in Ahmed Zewail's research group in Caltech from 1994 to 2007. He obtained his Ph.D. degree and worked in the Institute of Chemistry, Chinese Science Academy at Beijing, China from 1985 to 1989. He came to the United States as a postdoctoral scholar at the University of Kansas from 1989 to 1994.

25 Not Such an Important Man: He does not have Personal Security

Abderrazzak Douhal*

Hearing About Ahmed

The first time I heard about Ahmed Zewail was during my postdoctoral research at the Institute of Molecular Science at Okazaki, Japan, in 1990 while working in the group of Professor Keitaro Yoshihara. Professor Yoshihara used to attend the meetings on ultrafast science phenomena and related molecular science and he would share with us the latest news and advances in the field. Ahmed was among the important players and his scientific reports were often cited. I then became curious about the scientific results from this man with a classical Arabic name (Ahmed) working at Caltech. Before my postdoctoral work, I had been working in the group of Professor Ulises Acuña at the Institute of Physical Chemistry "Rocasolano" at the Spanish National Research Council, Madrid, on the spectroscopy and photophysics of new molecules showing photo-induced proton-transfer reactions in solutions. While at the Institute of Molecular Science, I had the chance to read Ahmed's paper on the spectroscopy of methyl salicylates (a classical molecule in the proton-transfer field) and its picosecond emission decays in a jet-cooled molecular beam. The vibrationally resolved blue

*Visiting Associate.
Email: Abderrazzak.Douhal@uclm.es

part of the dispersed emission spectrum attracted my attention. After Okazaki, I moved to CNRS (National Centre for Scientific Research)-University Paris-Sud to work in the group of another great scientist, Professor Françoise Lahmani. In search of the mechanism underlying proton motion, I carried out several experiments with Françoise and her colleague, Anne, on proton-transfer molecules in molecular beams. Hence, the research from Ahmed's group on methyl salicylate and other molecules depicting the concept of intramolecular vibrational energy redistribution of large molecules in molecular beams became increasingly interesting to me and relevant to my research. We later published a few papers on proton-transfer dyes and Françoise presented our results at a conference.

That was in 1993, and by that time I had already moved to Toledo (of Spain, not of Ohio) to work at the University of Castilla la Mancha (UCLM, Spain). In a phone discussion with Françoise, she remarked that Professor Ahmed Zewail liked the results that we had published on proton transfer in molecular beams. During the following two years at UCLM, my activities were mainly teaching and preparing new courses to deliver in the mornings and afternoons. At home, I used to analyse my research data from Japan and France and, when possible, carried out some experiments in my previous laboratory in Madrid. The lack of research facilities in Toledo at that time prompted me to look for more fruitful collaboration opportunities and ultimately a short stay in a world-known research group working on ultrafast spectroscopy.

Contacting Ahmed

At the end of 1994, I remembered Françoise's remarks regarding Ahmed's opinions about our work on proton-transfer, so I faxed Ahmed at the beginning of 1995 asking for a three-month stay during the summer to carry out femtosecond (fs) experiments using time-of-flight mass spectroscopy to elucidate the ultrafast proton-transfer mechanism in selected molecules. I have to say that Ahmed's response was surprisingly quick and, most importantly, he sounded very positive and excited. His enthusiasm was contagious as I then also became more excited about the experiments that I would conduct in his laboratory named Femtoland. I started to read many articles from Ahmed's group and got up to speed with how the technique worked and the concepts behind it.

I then wrote two short proposals and faxed them to Ahmed just as he had requested. One proposal was to determine if the double proton-transfer in the 7-azaindole (7AI) dimer occurred through a concerted or stepwise (1 + 1) mechanism. At the time, the spectroscopy of the 7AI dimer had already been studied in gas and solution phases, but there were no detailed studies on its mechanism. The photochemistry and related dynamics of the 7AI dimer are of photobiological interest particularly as the system mimics the hydrogen bonding of the DNA base pairs. The second proposal was to elucidate the proton-transfer mechanism of a molecule called HPPO which I had already studied in solutions while at Madrid and in molecular beams while at Paris. HPPO showed a dispersed emission spectrum and an energy dependence of the proton transfer reaction time in molecular beams.

Moving to Femtoland

Once I had sent both proposals and fixed a handful of issues related to my stay at Caltech, I moved for a three-month period at the beginning of June 1995. Ahmed Heikel, who was a Ph.D. student in Ahmed's group, was supposed to pick me up at Los Angeles airport. I cannot remember what happened but, possibly due to heavy traffic between Pasadena and Los Angeles airport, he was ultimately unable to arrive at the airport on time. I then called Ahmed asking if I should wait at the airport or proceed and go to his office. He gave me the name and address of a hotel in Pasadena and suggested that I take a taxi. The next day after my arrival (it was my first time in the United States), I met the great Mary, Ahmed's secretary at that time, who helped me arrange my stay at Caltech. I was very excited to meet Ahmed to talk about the planned experiments. After a short welcome meeting, he suggested that I finish taking care of all the administrative paperwork first before meeting with him the next day to discuss science. I remember that when I told him I wanted to talk about my experiments as soon as possible and that the administration issues (like opening a bank account and medical insurance) could wait, he laughed and asked Mary to help me and said (more or less), "He does not care about money!"

A few days later, we met to discuss the project in his office. I brought a thick and heavy dossier containing the most important papers on 7AI and some others on proton transfer in molecular beams which I had published with Françoise. With Arabic music playing softly in the background, Ahmed asked me to talk about my proposal. Ahmed inquired about what I will call the "intimate details" of the science presented in the papers. Of course, I could not give satisfactory responses to all his questions. Here, I will always remember this first major interaction I had with Ahmed. While having a profound discussion on 7AI dimer spectroscopy in molecular beams and condensed phase, I mentioned that the only published paper on 7AI in solution using fs-spectroscopy was one from a research group led by a person called Hochstrasser who was working in the United States, but that the authors did not consider what I was proposing: the nature of the double proton-transfer mechanism. However, I also mentioned that I did not know if Ahmed knew this research group. On hearing this, Ahmed laughed and then asked for more details on the paper. At that time, I did not know that Ahmed had been Professor Hochstrasser's Ph.D. student, and Ahmed did not reveal this fact during our chat. I only learned about it some days later, and one can only imagine how I felt after that. I should have found out more about what Ahmed had been doing before Caltech! This was my lesson learnt: Make sure that you have done your thorough due diligence on the subject as well as the person you are talking to. I promptly admitted my mistake to Ahmed afterwards.

After my whiteboard presentation in his office, he suggested that I choose just one system to work on since my stay of three months was relatively short. I decided to focus my research on 7AI and we met some days later with Dr. Sang Kyu Kim, a postdoctoral researcher in the group, to discuss the experimental plan. We had to wait for our turn to get the fs-setup free to use. Meanwhile, I suggested to Ahmed that I could work on writing a review article on proton-transfer dynamics while waiting for the laboratory to be available. He liked the idea and I was able to give him a draft to review in two weeks. He was surprised that it was prepared so fast. Some days later, he met to discuss the article and the parts that needed more clarity. That was my first real scientific interaction with Ahmed as I saw him in action making deep and precise comments. We submitted the review for publication in *Chemical Physics* since Ahmed had an invitation from the journal. I am so satisfied with that work as it is now one of the most cited papers on proton-transfer reaction dynamics.

You Do not Know How to Use the Vacuum Pump

Soon, we started the experiments on 7AI and I was impressed with Dr. Sang Kyu Kim's expertise in handling the system and resolving the technical problems. At this juncture, there is another funny anecdote which really impressed me. While doing the experiments for days and nights without stopping, taking turns and drinking lots of coffee to last through the night, the ultrahigh vacuum pump in the molecular beam setup broke down just as we were getting results. I ran to the first floor and asked Mary for a short and urgent meeting with Ahmed. Once in his office, I informed him about the bad news — I was especially sad after the initial excitement of getting the preliminary results! Then, with a large smile and his unique Egyptian humor, he said something along the lines of: "I see, you do not know how to use the vacuum pump, you just came from Morocco to break what we have here…!" I knew that he was only joking and trying to relax me. I then laughed and asked him what we should do now since we were getting exciting results. He told me to analyze the results with Sang Kyu and wait until the problem got resolved. The surprise was that we received a new pump in less than 48 hours and we could continue our experiments. Before Caltech, I had worked in several places, including Morocco, Spain, Japan, and France, but I had never experienced such efficiency in quickly resolving such problems. Some of the group members stated that this was what is was like working with Ahmed. I then remembered that Françoise would call it a "Zewail-type experiment" whenever we got nice experimental data in molecular-beam research.

We continued our experiments on the 7AI dimer and Ahmed called many times asking for updates on the experiments, even while travelling abroad for conferences. To me, these calls were a great help in enabling us to finish our plan in a couple of weeks as other group members were waiting for the setup. I also remember that Professor Joern Manz was visiting Ahmed's group and he gave a very nice talk on coherent control which included a wavepacket movie. He visited the sub-basement where we were running the experiments and was very interested in the 7AI results. Sang Kyu and I analyzed the results and, with Ahmed, we discussed the data and mechanism of double proton transfer in the 7AI dimer. Ahmed asked us to double check several experimental key points, the data analysis, and the presentation of the transients. After another meeting, I was tasked to write the first draft and within a few days it was ready. Ahmed gave me the corrections by hand, focusing on clarity and manuscript structure — the way to put together simple ideas so that they flow smoothly and reach clear and relevant conclusions. He also explained to me the reasons for the corrections he made. And here is another, third anecdote: I ran into Ahmed in the bathroom and he asked me, "Dr. Douhal (sometimes he called me "Zaiim Al-Oroba," which in Arabic means "Leader of Arabism"), how do you like my suggestions and corrections of the draft?" I jokingly said, "I prefer my version." He then laughed and retorted, "Oh, if you want, we can send it to ... instead of *Nature*." Of course, he knew that I was kidding; it was my little revenge on his joke when the vacuum pump broke. I can still recall this conversation as if it had happened yesterday and it shows how Ahmed loved science, even in the bathroom. We sent the paper to *Nature*, a top scientific journal, and it was accepted after few weeks. I was very satisfied with the fact that, in my short stay of three months, I managed to produce a review on proton-transfer reaction dynamics and publish an article in *Nature*. Ahmed was also very pleased, later confessing that he did not expect all of that from me in such a short time. That was also a solid basis to continue my collaboration with Ahmed for more than six years while doing what I could from my modest steady-state spectroscopy laboratory in Toledo. I also had the pleasure of working with Torsten and

Mirianas, two fantastic postdoctoral researchers, in elucidating the dynamics of 7AI dimers in solution.

There is another amusing anecdote of Ahmed's humor when I visited Femtoland for the second time in 1998 to work with Torsten and Mirianas on new experiments. While preparing for a group seminar, I was looking at the printed slides in my Femtoland office and Ahmed approached without my noticing him. Once behind me, he asked, "Abderrazzak, what is the breaking news?" The first slide of my talk was a big black and white picture of my son Yunas, who was then only a few months old. I took it out and said, "This is the most important result of 1997." He laughed and then said, "You are right, it is a very important result!" Of course, we quickly got down to business and I then showed him the other slides and briefly mentioned what I wanted to talk about for the seminar.

Later on, Dongping, a Ph.D. student in Ahmed's group, and I examined the dynamics of a molecule (HPMO) that is comparable in structure to the one I had proposed to investigate in 1995. With my research group in Toledo, we had published a paper on its spectroscopy both when in solution and when trapped in chemical and biological cavities. Ahmed liked the idea of elucidating the dynamics of proton transfer within these cavities as he wanted to continue with the concept that we had already examined using iodine molecules in cyclodextrins. I moved to Caltech for a couple of weeks to carry out fs-experiments with Dongping and, after a few months, we published the results in the *Proceedings of the National Academy of Sciences*. I must say that Ahmed was generous enough to make clear in his Nobel paper that the work in caging media (micelles, cyclodextrins, and human serum albumin protein) was the result of a collaboration with me.

After the Nobel Prize Committee announcement of the award for chemistry in 1999, I was very happy to know that this most prestigious award was for Ahmed, with whom I had collaborated closely, in addition to him being the first Arab to get a Nobel Prize in science. My satisfaction and feelings were immense. Ahmed invited me for the event organized at Caltech in January 2000. Many of his friends and colleagues were there and they thoroughly enjoyed the fiesta

Photo 25.1. From left to right, and top to bottom. In the femto lab doing experiments using femtosecond spectroscopy (Caltech, July 1995). Ahmed with a few members of his scientific family at the FemtoV Conference (UCLM, Toledo, September 2001). Dongping Zhong, myself, Ahmed and Dongping´s wife (Caltech, January 2000). With Spencer at observatory in Los Angeles (August 1995). Ahmed with my group´s members and colleagues at FemtoV Conference (UCLM, Toledo, September 2001). Celebrating Nobel Prize award at Caltech (January 2000). With Ahmed in Toledo (May 2001). With Ahmed, Yasmin and Yunas at FemtoV (UCLM, Toledo, September 2001). With Ahmed´s group members at Caltech (July 1995). With Spencer in Pasadena (August 1995).

(see Photo 25.1). Here, I have another anecdote of how much importance Ahmed placed on the small details. Just before entering the Athenaeum at Caltech where the celebration of the Nobel Prize award was to be held, I met Ahmed again and, after shaking hands, he looked at me, took my tie, and straightened it out as it was not well fastened (see Photo 25.1). This little act exemplifies his detail-oriented and attentive personality; whereas I was simply too excited and enjoying the unique moment, Ahmed was still well aware of everything around him. I was certainly very honored and touched when Ahmed called me to his office and sought my opinions on the figures that he wanted to use as he applied his finishing touches to the Nobel paper.

Femtochemistry Conference in Toledo and Visit to UCLM

Within the first week of September 2001, I organized the Femto V conference with Professor Jesus Santamaria at the UCLM campus in Toledo. One year before the conference took place, I met Ahmed during a conference in Switzerland and discussed with him how the scientific program should be set with the members of the international committee. Ahmed suggested a few names and asked me to contact the other committee members. Ahmed insisted also in sending clear letters to the invited speakers, indicating how much the event will cost and what the local organizers will cover. This astute piece of advice prevented arguments from arising with some speakers who wanted more financial support after accepting the invitation. At the time of the conference in Toledo, the weather was really hot (around 40°C, but dry), and the day before Ahmed's lecture, he asked me to turn the air-conditioner on as early as possible to cool the room. The morning session went very well, but for the afternoon session, the temperature started to increase and I had to ask the technician to fix the air-conditioner so that the temperature was more bearable for the attendees. Ahmed, with the other members of the international committee of Femto V, were very satisfied with the arrangements and smooth ambience of the conference. Femto V, with more than 250 participants from abroad (see Photo 25.1.), was a special event for Ahmed as well as for many people, as they were looking for new advances and directions for the community after the Nobel Prize was awarded to the field. Ahmed nicely presented the state-of-the-art research in femtochemistry and femtobiology and acquired promising experimental results using a new ultrafast electron diffraction setup. Other talks also showed fantastic contributions in theory and experiments.

Before the Femto V conference, Ahmed graciously accepted my invitation to visit our Toledo campus and to deliver a public conference in May 2000 (see Photo 25.1). The conference room was so full that some people could not even get a seat, but everyone present enjoyed the lecture. Even high school students came to the event. The lecture was wonderful and Ahmed was very happy to see many young science enthusiasts. Numerous colleagues at the university congratulated and thanked me for bringing Ahmed as they appreciated the opportunity to listen to such a lecture on the most advanced frontiers of science, remarking on Ahmed's incredible ability in making the science clear and easy to follow.

I received similar comments when Ahmed came for the third time to Toledo to give the opening lecture of the International Conference on Photochemistry in 2009. We hosted Ahmed in a small but pretty and classical hotel on the other side of the Tajo river surrounding Toledo. After delivering the Nobel lecture on the Toledo campus, I invited Ahmed to the district where we live. While walking to dinner at a terrace with my family, I said to my son Yunas, who was four-years old at that time, that this man walking with us was one of the most important people in the world. With the complete innocence of a four-year-old kid, Yunas asked in Spanish, "If he is such an important man, why does he not have personal security?" After translating my son's remark for Ahmed, he laughed dearly and said, "He *is* smart; I do not have personal security, right, I am not so important." This anecdote is linked with another a few years later in Egypt.

While attending the Modern Trends in Physics Research conference organized by Professor Lotfia Nadi in 2006 in El-Cairo, Egypt's Minister of Education and Science at that time warmly welcomed the participants and Ahmed during the opening ceremony and said, "We have an Egyptian saying: if you drink water from the Nile river, you will come again to Egypt." Later, before Ahmed started his lecture, he thanked the Minister, welcomed the participants, and said, especially to the foreigners in a very polite and humorous way, "I am sure about the great hospitality of Egypt, and of the Minister of Education and Science. But please, do not try to drink water from the Nile. We want you to have a nice stay, wonderful reminiscences, and a safe trip back to your homeland so that we can have you back again in Egypt next time." Of course, you can imagine that all the people at the conference room laughed and enjoyed Ahmed's comment, including the Minister! And here is the second anecdote in Egypt, which relates to Yunas's comment a few years before in Toledo. While enjoying a very nice evening at the famous Khan Alkhalili area in El-Cairo, Ahmed invited us and other friends attending the conference to another beautiful place. While on the way there, I realized that personal security men were walking in front and behind us. People from Egypt easily recognized Ahmed and wanted to show their respect to him and shake his hand. Even before the Nobel award, Egypt had already commemorated Ahmed with his image on an official post-office stamp and many television programs had been devoted to his education in Egypt and his research in the United States. So, once we were seated and served at the restaurant, I said to Ahmed, "It is a pity that Yunas is not here to see the personal security men and be convinced that you are an important man." He laughed heartily, remarking once again that he was a smart kid. We enjoyed a very nice evening filled with local drinks and sweets. Ahmed was very happy to have so many friends from Europe in his homeland.

International Conference on Photochemistry, ICP09

In July 2009, I organized the International Conference on Photochemistry (ICP09) in Toledo. More than 550 participants from more than 33 countries gathered. Ahmed delivered the opening lecture and showed new exciting advances using ultrafast electron diffraction and microscopy. After his talk, many people thanked him for the wonderful talk and new scientific findings, and Ahmed was pleased to hear that so many people were interested in his groundbreaking work. Overall, we enjoyed a wonderful ICP09 conference with many important speakers, and the participants also enjoyed Toledo and her food, tricultural heritage, evenings when the weather got more enjoyable, and magical atmosphere at night.

We took a conference picture during the afternoon and, unfortunately, Ahmed was relaxing in his room at that time and could not be in the picture. Upon hearing that he was left out of the group photograph, he asked me why I had not called him and then jokingly suggested, "You can photoshop me in!" At ICP09, Ahmed gave an excellent interview to a freelance journalist with an emphasis on the future and efforts to make advancements in basic science, including the development of new technologies and education. The interview was published in *El País*, the most important newspaper in Spain. Before planning this visit to Toledo, Ahmed's secretary had asked me for details such as my work and home addresses, my office and home phone numbers, and the hotel's full address. I was surprised that they needed all this information and jokingly assured that he will be in the safest city in the world and also in good hands. The secretary laughed dearly. All of Ahmed's secretaries with whom I had the pleasure of interacting with were of the same calibre as Ahmed's spirit — responsible but also good humored. They only needed the information in one document because it was required by the security office at Caltech, as Ahmed was a member of the scientific advisory board of the President of the United States, Barack Obama.

Ahmed's Last Visit to Toledo, July 2015

Ahmed came back to Toledo for the fourth time in the middle of July 2015 to deliver the opening lecture at an international conference organized by a colleague from Madrid. Dema was with him and Luis Bañares was the host. Ahmed once again showed his new advances in ultrafast electron microscopy. Although Ahmed had been ill for the past two years, he was all ready to travel and talk about science and technology at the conference. After his lecture and the press media interview with Luis, we went for lunch at a restaurant on the other side of the river which boasted a gorgeous view of Toledo. After a short, relaxing time at the hotel lobby, Luis drove us under the July sun to the restaurant. Ahmed asked me, "What's new, Dr. Douhal?" I answered that, with the help of Villy's group, we had built a fs-setup to carry out time-resolved THz experiments on photoconductive materials such as perovskites, quantum dots, and others. He said it was important to build new techniques and to explore new phenomena. During lunch, we talked about science in Spain and the sociopolitical situation in the Arab world. Once he had finished his lunch, he told us the following (a paraphrasing on my part), which would (unfortunately) be the last anecdote that I can share: Upon finishing a lecture in a conference, a participant approached Ahmed, thanked him for the lecture, and said, "Professor Zewail, you delivered a clear and nice talk, and you have such a wonderful and deep voice. You should be a politician for the good of the people, and not like those of today; they do not know what they are saying."

Before the summer of 2016, I contacted Ahmed to ask if we could collaborate and write a proposal on smart new materials for the 2016 call on scientific collaboration between the United States and Spain. The idea was to explore fs-dynamics in Toledo using the available techniques in my research group (after two decades, we finally have our own fs-labs in Toledo!) and to elucidate other phenomena with single crystals using the ultrafast electron microscopy at Caltech. I did not know that Ahmed was ill again and rather seriously this time. Unfortunately, on the night of 3 August, I got an urgent mail from a medical doctor (whom I did not personally know) informing me about Ahmed's death and giving me his condolences. Obviously, when receiving such tragic and unimaginable news, you feel compelled to check and double check whether it is true. I called Luis but there was no response; he was deep in the depths of these sad moments at Caltech and was too distraught to respond to phone calls. I then called Majid and he confirmed the bad news. Mails started to pour in from many friends and colleagues of the femtochemistry community. Those were tremendously sad days.

I decided to make a book to reminiscence about Ahmed and his scientific family as much as I could. I then talked to Majed, who informed me that there is indeed a book being written, although different from the one that I had conceived of. I then suggested the idea to Dongping and Spencer, who were both still shocked and saddened by the news. They agreed with the idea and, a few weeks later, we started to work on it.

At the end of telling my story and reminiscences of Ahmed, I have to say that our relationship, in addition to being a strong scientific collaboration, was friendly and respectful. I used to call him "Ostad Ahmed," which in Arabic means "Professor Ahmed," but also carries with it the highest level of consideration and respect possible. He used to call me "Zaim Al-oroba" (when joking), "Abderrazzak" (the Egyptian way), or "Dr. Douhal" (the American way) when talking about science. Before 2009, we wanted to arrange for Ahmed to visit my homeland, Morocco. However, at that time, he became involved in President Obama's government and I came to the realization that organizing this visit by myself would have been too complicated and beyond my means, as it would have involved complex procedures of diplomacy between the two countries, Morocco and the United States. During his last visit to Toledo in July 2015, when Ahmed was enquiring about science in Morocco, he commented that he has been in various Arabic countries to give scientific and public educational lectures but not Morocco. I told him about the aforementioned plan and why I could not make it happen, but I also promised to work on it the following year. However, and unfortunately, he could not spare the time.

Ahmed lived an intense life that was spent elucidating the short events of chemical reactions and their mechanisms which shape the "birth and life of molecules." He directly trained hundreds of scientists and made a great impact on thousands of others. He was a great man, a great scientist, a great thinker, and, for me, a great friend. The world has lost an important man but his strong legacy of science, technology, and culture lives on.

Abderrazzak Douhal was born in Beni Mellal, Morocco, and he received his Ph.D. degree in chemistry (physical chemistry) from Kadi Ayyad University, Marrakech, after a period of doing research at the Faculty of Sciences at the same university and at the Institute of Physical Chemistry "Rocasolano" at the Consejo Superior de Investigaciones Cientificas in Madrid. He is a professor of physical chemistry at the University of Castilla La Mancha (UCLM), Toledo, Spain. Between 1990 to 1992, he was a postdoctoral researcher at the Institute for Molecular Science in Okazaki, Japan. In 1993, he worked as a research associate at the Laboratoire de Photophysique Moléculaire of CNRS (National Centre for Scientific Research)-University Paris-Sud, France. He was a visiting researcher at California Institute of Technology several times between 1995 and 2000 to collaborate with Professor Ahmed H. Zewail. Since 1998, he has headed the Femtoscience and Microscopy research group at UCLM, focusing his research on the study of ultrafast photoevents in condensed phases and advanced materials.

26 The Affinity between Me and Ahmed Zewail

Eric Wei-Guang Diau*

My first brush with Ahmed Zewail's (AZ) research was during my senior year of undergraduate study in 1984 when I read his paper published in 1980 in *Physics Today*, titled "Laser selective chemistry — is it possible?" (number 11 of volume 33, pages 25 to 33). I wrote a report for the class I took based on the concept that AZ provided in that article. Later, I entered graduate school in National Tsing Hua University, Hsinchu, Taiwan, and completed my Ph.D. study in 1991 in the field of reaction kinetics using the laser-induced fluorescence technique to measure OH kinetics in atmospheric chemistry and combustion chemistry. After two years of military service, I wrote to AZ regarding the possibility of being a postdoctoral fellow in his group in 1993. The first attempt was unsuccessful because the number of publications I had was quite limited, so I joined another group in the US in 1993. My second attempt to contact AZ was in 1996 when I was in Australia for my second postdoctoral term. AZ replied promptly and positively and sent me an application form to fill out in order to be a postdoctoral fellow at Caltech. That was really good news for me. Unfortunately, the offer from AZ did not come on time when I finished my term in Australia in the summer of 1996. I therefore went instead to Stanford Research Institute (SRI) International in the Bay area in October that year.

* Postdoctural Research Fellow.
Email: diau@mail.nctu.edu.tw

I still remember that phone call from AZ to my office at the SRI International. The time was 8.00 p.m. on 24 December 1996 and he wanted me to come to Caltech as soon as possible although I had a one-year contract with SRI. I made a quick decision to join AZ's group the following spring and paid the penalty for breaking the contract and leaving SRI International earlier. As a result, I joined AZ's group on April 1, 1997 and found myself working in Femtoland in the sub-basement of Noyes building. I had worked with nanosecond YAG lasers before, but that was my first time working with a femtosecond laser system, in particular the first-generation CPM laser which was pumped by a YAG-dye laser system to generate the fs pulse at 620 nm. This 620 nm fs pulse was then doubled using second harmonic generation to produce the fs pulse at 310 nm as a pump, which was then used to excite small organic molecules. The generated intermediates were probed by time-of-flight mass spectrometry using the 620-nm fs laser pulse via a multi-photon ionisation procedure. In this way, we managed to study many organic systems, including ketones, cycloketones, ethers, azulene, azomethane, alkyl bromides, and many others during the 1997 to 2001 period of my stay in Caltech.

During my stay in AZ's group, my impression of AZ, which I think every other group member will also agree, is that he is a very busy person — he traveled a lot so you wouldn't see him very often. The first year for me was particularly tough because I did not know his style — he did not like "half-cooked" results and drafts — so I couldn't simply show him my daily results. Unless you really got breakthrough results and you wanted him to be the first to know, it was otherwise very difficult to make an appointment with AZ due to his tight schedule. So what I needed to do was to finish the experiments, carry out the calculations, and figure out a reasonable mechanism to explain the experimental and calculated results, before I could even dream of getting in touch with AZ. But even that was not enough; you would have to finish writing up a paper that was ready to go first! Then you could give the paper to AZ, go back to the laboratory and work on another project, and wait for his phone call. Usually, he would read my drafts during his travel. When the results impressed him, the phone call would come after he returned from the trip. The way he worked out a paper was very efficient. He would start reading my paper together with me, ask me questions, and make corrections right away. In a couple of hours, the editing of the paper would be completed and the paper would be ready for submission.

AZ had many great ideas in his mind, but the chance to talk to him in person was scarce. Such opportunities presented themselves whenever we finished a paper together and he would invite me for a cup of coffee. Of course, during those rare opportunities we would talk about things beyond science, but that only ever happened twice in my memory. The most exciting time was when I was involved in the big events of 1999 when AZ received the Nobel Prize in chemistry. Many events and gatherings were organized to celebrate this huge triumph and I was lucky to be involved in those big celebrations and had the chance to take a photo with AZ as shown in Photo 26.1.

Photo 26.1. Ahmed Zewail and I during the celebration of his Nobel Prize in Chemistry.

Photo 26.2. Group photograph including me, Ahmed Zewail, and many other group members in front of the El Portal restaurant before my departure from Caltech in 2001.

After working in AZ's group for more than four years, I finally found a good academic position in Taiwan with AZ's strong endorsement. Before leaving Caltech in 2001, he kindly arranged a group dinner at a Mexican restaurant and the whole group took a photograph together afterwards (Photo 26.2). I still remember that AZ was very happy during the dinner and he punched my shoulder and said, "From now on you can call me Ahmed!"

After I returned to Taiwan, I never had the chance to meet AZ again until eight years later. I found out that our university, National Chiao Tung University (NCTU), which belonged to the United University System of Taiwan, had a program which supported the visit of Nobel laureates to Taiwan using funds provided by the Sayling Wen Cultural and Education Foundation of Taiwan. As a result, I wrote to AZ regarding this opportunity to visit Taiwan and that NCTU will also honor him with an honorary doctorate degree. After receiving this message, AZ called me right away to fix his schedule

Photo 26.3. Picture taken in 2010 of Ahmed Zewail and all his former Taiwanese postdoctoral researchers. From left to right: James (Po-Yuan) Cheng, me (Eric Diau), Ahmed Zewail, Jau Tang, and Juen-Kai Wang.

for a visit to Taiwan. The visit happened June 25 to June 30, 2010 and AZ gave three lectures — one in NCTU, which was more general for his honorary doctorate degree, and the other two at the National Yang-Min University and Academia Sinica. In particular for the lecture at Academia Sinica, AZ met all his former postdoctoral researchers then based in Taiwan (Photo 26.3). Although AZ's visit was short, his lectures inspired many researchers in Taiwan. For the last day of his visit, James and I sent him to the airport and we chatted about the scientific developments in Taiwan. We were hoping that he could visit Taiwan again to stay for a longer period of time and get a sense of why the Portuguese called it "Ilha Formosa" (Beautiful Island) in the 16th century. However, this hope diminished in August 2016, when I heard the bad news about AZ.

My stay in AZ's group lasted four years and four months in total. What I learned from AZ is the spirit of science which he summed up as "CACIS" — curiosity, accuracy, consistency, insistency, and simplicity. He once told me that the purpose of science is to make things simple rather than complicated. He was always searching for the simplest solution to solve a complicated problem in science. Even though he had passed away, his scientific spirit will be alive forever! I published 16 papers with AZ and his group during my stay in Caltech, and we managed one more publication after I returned to Taiwan to start my own academic career at NCTU. I would like to thank all the co-workers who toiled hard with me during the time I was in Caltech and for all the good times; they will never come back again but they will always be etched indelibly in my memory for life! Thank you Ahmed Zewail; the world is a better place now because of your tremendous efforts both in science as well as beyond science!

Eric Wei-Guang Diau received his Ph.D. in physical chemistry from National Tsing Hua University, Taiwan, in 1991. Before joining the Department of Applied Chemistry at National Chiao Tung University, Hsinchu, Taiwan, as a faculty member in 2001, he worked as a postdoctoral fellow at Emory University (1993–1995), University of Queensland (1995–1996), Stanford Research Institute International (1996–1997), and California Institute of Technology (1997–2001). He is interested in studying the relaxation kinetics of condensed matter, in particular interfacial electron transfer and energy transfer dynamics in many light-harvesting and light-emitting systems. His current research focuses on the developments of novel functional materials for next-generation solar cells and light-emitting diodes, including perovskite solar cells and micro-light-emitting diodes. He received the Outstanding Research Award at the Materials Research Society Spring Meeting and Exhibit in April 2014, and the Sun Yat Sen Academic Award from the Sun Yat Sen Academic and Cultural Foundation in October 2014. He has published over 180 peer-reviewed papers and has an H-index of 52 till the end of 2017. He is currently a Distinguished Professor at the Department of Applied Chemistry and Institute of Molecular Science at National Chiao Tung University.

27 Ahmed Zewail — The Making of a Man and Great Scientist

Torsten Fiebig*

Like most of Ahmed's students and postdoctoral researchers in the late 1990s, I was fascinated with the personal "mystique" that surrounded Ahmed Zewail. Before joining his group at Caltech, I had heard many rumors about this man, such as his steadfast aspiration for the Nobel Prize and the soaring intensity of his group. Prior to my arrival in March 1997, I had met Ahmed only once during the 1996 American Chemical Society Meeting in New Orleans where he gave an animated keynote lecture. That lecture fueled the excitement of many young scientists — myself included — not just for ultrafast spectroscopy, but for *something else* as well. If you were young in the 1990s and had a sharp mind that was easily excitable by thoughts about complex molecular processes and quantum optics, there were quite a few outstanding academic places to choose from, in the state of California alone. Hence, joining Ahmed Zewail's group at Caltech was by no means an "inevitability" in one's personal career trajectory. After all, some of the greatest minds who have shaped physics, chemistry, and biology over the past 100 years were/are rooted in this part of the world. However, I admit that the experience of meeting "Dr. Zewail" in 1996 was profound for me in the sense that I wanted to learn more about this *something else* that connected the person Ahmed Zewail to the mysterious nature of

* Postdoctoral Research Fellow.
Email: torsten.fiebig@gmail.com

chemistry and physics. In short, I wanted to understand what it was that made him such a prophet-like figure in science, someone who possessed the talent of generating great enthusiasm and excitement among a large group of disciples.

Peeling the Onion

The famous psychotherapist Fritz Perls used to phenomenologically describe the human psyche in terms of distinct layers (the "Onion model"), such as the cliché, role, phobic, and core. Each layer has its place in our everyday lives, both in terms of normal and abnormal functioning. Gaining a (nearly) complete understanding of someone requires us to peel the onion layer by layer. It is interesting to note that Ahmed enjoyed playing a vaguely related game himself on many occasions, usually after group dinners or at social gatherings over the holidays, when he and his wife Dema would invite students and postdoctoral researchers who did not have family members living close by. On such occasions, we would discuss the layers of our personality (at least those that would represent our scientific ambitions!). Looking back at the layers that Ahmed revealed about himself, it was clear and apparent to me that he possessed a childlike imagination and a profound curiosity, combined with an unquenchable desire to understand how nature works on its most fundamental levels; that is, the time and space scales of atomic motions. While these qualities are certainly well-suited for success in any academic career, they seemed hardly sufficient to lead the purpose-driven life that Ahmed had pursued on his journey through life.

Picking the Truly Important Problems

As a young assistant professor at Caltech, Ahmed started out studying atomic motions in small, isolated molecules in the gas phase. The advent of ultrashort laser pulses in the 1970s opened up fascinating new experimental avenues for exploring molecular dynamics in real time. No one recognized and foresaw the disrupting effects that ultrafast laser technology would have on science in general, and on chemistry in particular, more clearly than Ahmed Zewail. Lasers would not simply be another tool to measure fast chemical kinetics; instead, they would enable scientists to generate (possibly manipulate!) and monitor coherent atomic motions in real time based on the physical "blueprint" derived for nuclear magnetic resonance a couple of decades earlier. Through a series of pioneering experiments, Zewail succeeded in translating his intuition-fueled vision into solid groundbreaking experimental results that would ultimately become integral parts of future chemistry textbooks. He and his co-workers visualized the rotational and vibrational atomic wave packets, the flow of energy within molecules, and the making and breaking of chemical bonds. Concepts that were hitherto only theoretical ideas became experimentally proven and formed the pillars of a new field called "femtochemistry."

The way in which Zewail responded to his early success turned out to be another critical component of this man's extraordinary making. After femtochemistry had become an emergent research field, there were many new experimental variations and ideas that could have provided Zewail with enough intellectual material to keep his "Femtolands" running well into his retirement. Continuing to invest his resources, expertise, and efforts in this emergent field of science would have been what most scientists in similar situations would likely have chosen to do. After all, Zewail's work had raised a number of important and fundamental new scientific questions which could hardly be characterized as "low hanging fruit."

However, it was not in Ahmed's nature to undertake small steps that, from a distance, some might misinterpret as standing still. I learned that lesson early on during my first project in Ahmed's group. There had been a lively debate in the literature about the mechanism of two-proton tautomerisms in N-heterocyclic base pair analogs where symmetry would, in principle, allow concerted, even synchronous double proton transfer after photoexcitation. One system with such properties is 7-azaindole (7-AI). In 1995, Abderrazzak Douhal, Sang Kyu Kim, and Ahmed published the gas phase dynamics of 7-AI dimers which indicated a stepwise, sequential two-proton transfer, in a letter to *Nature*. When I came to Caltech, Zewail assigned Mirianas Chachisvilis and me to carry out a similar study in organic solvents. I was fascinated by this project (at least at first) because it would enable us to directly determine the effect of the solvent by comparing femtosecond fluorescence measurements with results obtained in the gas phase. However, my excitement was rather short-lived when we found out how enormously difficult these experiments really were. I had worked with numerous organic materials that would undergo photoinduced deterioration, but I'd never seen a compound as unstable as 7-AI. Not only does 7-AI decay very quickly upon laser irradiation, but in the presence of residual air dissolved in the solvent, it forms sticky semitransparent films on the cuvette surface and reacts with metal parts inside the flow cell pumps. Finally, we succeeded by designing and customizing an experimental setup that was uniquely capable of handling 7-AI under various solvent or temperature conditions.

By the time we were ready to show our results to Ahmed, my friend Mirianas and I had had dozens of discussions about possible mechanisms and, given the relevance of 7-AI dimers as DNA base pair analogs, we both felt that further studies on 7-AI could provide more mechanistic insight into DNA photophysics and photoinduced mutagenesis. However, soon after Mirianas and I had presented our plans for future experiments to him, Ahmed simply shrugged and said, "Gentlemen, let me tell you something. You have just demonstrated beautifully what was clear to me from the very beginning. There is no way that the protons will move in a concerted fashion along the potential energy surface." At first, I was very disappointed. After having just spent several months on the nuts and bolts of solving real experimental problems, I thought it made no sense to stop at this point now that we had come so close to producing at least two or three more papers on the subject. The next day, I met Ahmed in the hallway and he said, "Torsten, I understand you want to go deeper, study different solvents, and vary temperature; all of this is reasonable and believe me, others will do that. They will work on this problem for decades to come. But I just simply don't have the time. We were going to show that the mechanism is stepwise and that's it. Let others vary the temperature and all of that stuff. You have to learn that the day has only 24 hours and that life is limited. There are so many important problems out there. I have more ideas for great experiments than I will ever have time to do." To some of you who knew Ahmed well, those words will sound familiar. When I heard them first in 1997, my entire outlook on science changed. In fact, I remember leaving early that day from the laboratory to write these words down and think about them, as they may reflect the core of Ahmed Zewail and the secret of his success better than anything I had heard before.

Anyone as ambitious and driven as Ahmed will inevitably get frustrated on occasions when his or her high expectations are not met by the people he or she is working with. In situations like these, Ahmed frequently resorted to his very unique style of humor mixed with sarcasm as a conversational tool to express criticism and even frustrations. Interestingly enough, his often witty and funny remarks were not exclusively reserved for group members but sometimes even directed at other prestigious Caltech faculty members. I vividly recall a lunch with him and my team mates for the "DNA electron transfer" project in late 1998. We were discussing, quite controversially, a myriad of possible theoretical models that could explain our experimental data when Nobel laureate Rudy Marcus entered the room. The next thing I remember is Rudy staring at the whiteboard and then making a general comment about activation barriers in DNA-mediated transfer processes. Ahmed, still heated up from the discussion, quipped, "Rudy, these are very complex biological systems, not just two parabolas shifted around on paper."

Strong Confidence in his Own Vision

About a month after the Nobel Award Ceremony in Stockholm, we had a group seminar in Ahmed's conference room where he shared personal stories and memories about the event with us. Strangely, after working for almost three years in Zewail's group, this incredible Prize had been omnipresent during every discussion between students, postdoctoral researchers, and in meetings with Ahmed as well (naturally!). Although it hardly ever was the subject of real conversation, the Nobel Prize did implicitly play a key role in defining *who we were* as young scientists, and, maybe more importantly, now that this statistically highly unlikely event had actually occurred, where would the group go from here?

Finally, Ahmed asked us if we had any questions regarding the royal procedures or the Nobel Prize in general. I decided to ask him two questions. "Ahmed," I asked (at that time I had officially matured to call him by his first name, even in front of the whole group… or so I thought at least), "we all know that you have been waiting for this Prize to come. Was there anything different this year? Did you get any senses or signals that it might really happen?" He smiled during the first part of my question and then went on to assure us that he had had no idea. My second question may have been more daring and several group members asked me afterwards why I had asked it. "Ahmed, were you surprised that you did not share the Prize with anyone? After all, most of the Nobel Prizes are now awarded to groups of two or three scientists." I still wanted to go on and further elaborate on my point, but his facial expression told me to pause. Yes, Ahmed got almost defensive and, although he did not seem upset at me, it became apparent that he simply did not understand why I had asked this question. He said, "I think the Nobel Committee was very clear about that point. They praised my work, in particular the founding of the field femtochemistry! Nobody else has done that."

Ahmed gave several media interviews in which he highlighted the unusualness of his young age (he was only 53) for anyone to receive the Prize. However, he emphasized that this fact was not simply a curiosity but rather a proclamation to the world that his greatest scientific accomplishments had hitherto not yet been made. A few months later, he repeated this statement during the Nobel Prize Award Ceremony. But his Nobel lecture in front of his Majesty Carl XVI Gustaf, the King of Sweden, contained another remarkable phrase that would render many friends and foes (both scientists and non-scientists) speechless around the world. The comment was made during the last part of his lecture where he speculated about his own future as a scientist. Paraphrased, it said, "I looked at the Award Laudatio by the Nobel Committee very carefully, and I was very pleased to see that our most recent work on electron diffraction had not been mentioned in it. That fact gives me hope that we might see each other here again someday in the future."

I strongly recommend everyone to watch the entire lecture online. It is a fascinating account of Ahmed's scientific worldviews and how they are linked to him and his work: http://www.nobelprize.org/nobel_prizes/chemistry/laureates/1999/zewail-lecture.html

Kindness at the Core

When it was finally time for me to leave Caltech and move on with my scientific career, Ahmed invited me to a nice lunch in one of Pasadena's fanciest Italian restaurants. We had an extensive and lively discussion about life, science, and the future. Afterwards, he treated me to a premium Cuban cigar. It was a very special and intimate moment. We both reflected on the past three years. Gone was his sarcastic humor; instead, he spoke from the heart. I felt that our relationship had undergone a powerful transformation. Now, almost 20 years later, it is this kindness under a demanding and sometimes rough surface that I remember most vividly when I think of Ahmed and my wonderful time at Caltech.

Photo 27.1. Ahmed and I enjoying a nice cigar after the Nobel Prize celebration dinner at the Atheneum on January 15, 2000.

Torsten Fiebig graduated from the University of Göttingen, Germany in 1993 and received his Ph.D. degree in chemistry from the Max Planck Institute for Biophysical Chemistry in 1996. From 1997 to 2000, Torsten was a postdoctoral research fellow at the California Institute of Technology, working with Professor Ahmed H. Zewail. After three years as an Emmy-Noether faculty member at the Technical University of Munich, Germany, Torsten received his habilitation (equivalent to a D.Sc.) in chemistry in 2004, after having joined the faculty at Boston College, Boston, Massachusetts in 2002. In 2010, Torsten accepted a dual appointment as Research Associate Professor in the Chemistry Department at Northwestern University, Evanston, Illinois, and Director of Operation of a Department-of-Energy funded multicenter energy frontier research center (EFRC). In 2017, Torsten joined the Beckman Laser Institute and Medical Clinic at the University of California, Irvine as an entrepreneur to incubate a company in medical laser therapeutics. His research interests include medical photonics and laser-tissue interactions.

28 Personal Reminiscences of Ahmed H. Zewail

Thorsten M. Bernhardt*

Two weeks after the defense of my Ph.D. thesis in Berlin, Germany, I arrived at Caltech to start my postdoctoral position with Ahmed Zewail. This was in December 1997. I stayed for two years at the Noyes laboratory and returned again several times in 2000 and 2001 to work in laboratory 048 in the sub-basement of Noyes, and to help complete the new molecular ion beam machine that we started building in my second year as a postdoctoral researcher together with Hern Paik.

During my first year, I was fortunate to be able to collaborate with Dongping Zhong in the same laboratory. From him, I received my primer in time-resolved laser spectroscopy as I did my Ph.D. in molecular beam science but did not use any lasers. Accepting a complete newcomer in a world-renowned laser spectroscopy group is something I will forever be grateful for to AZ. I believe that there are more than a few postdoctoral researchers of AZ who would feel the same gratitude. This clearly demonstrates his trust in the abilities of and his support for young scientists who were captivated by the excitement of his research.

It probably has already been mentioned by other contributors that most of us used to call Ahmed Zewail "AZ" (when not directly addressing him). This was because he had the habit of marking the scientific journals that he had already skimmed through with his

* Postdoctoral Research Fellow.
Email: thorsten.bernhardt@uni-ulm.de

initials AZ before passing them down to our coffee corner in the sub-basement laboratories. I never dared to address him in a more personal way than "Dr. Zewail" unlike some of the others, but this is perhaps as much due to my respect for him as it is to some quite German conventions. However, I like the more familiar "AZ" when talking about Ahmed Zewail and therefore I hope I might be granted to use it throughout these brief reminiscences.

Certainly, there is a good chance that some or even all of my personal experiences with AZ are quite similar to those of my fellow collaborators. Nevertheless, I will dare to share those memories of AZ that still influence my work today and that I, from time to time, pass forward to my students.

Some experiences made AZ very special as a great scientist to me personally. When I was very new to the laboratory, Dongping used to start the experiment and sometimes left me to guard the data acquisition. AZ had the habit of coming down to do a tour of the sub-basement laboratories around 6.00 p.m. at those times. So it happened that I was alone in the laboratory recording the data when he entered. He looked at me and then at the computer screen where the newly acquired transient data was displayed. He immediately told me that there must be something wrong and that I should start the data acquisition at earlier delay times and not miss the complete time zero information for the later fitting procedure. Of course, he was right and I was ashamed of my ignorance, but I was also tremendously amazed by his interest in the most intricate details of the research that was going on in his laboratories. Later on, I really looked forward to those 6.00 p.m. laboratory visits of his because they also offered a chance for me to discuss with him our daily hands-on laboratory and science issues. I took over this habit when I started my own group and I still try to pay my laboratories and the students doing research there a visit at least once a day.

Before starting my postdoctoral research with AZ at Caltech, I met him only once at a conference. He was invited as a plenary speaker and I admired the way he presented his research. Everything I gathered from his published papers also appeared to me so perfectly conclusive and convincing. After all, this was shortly before 1999 and almost everybody agreed that he had a good chance of receiving the Nobel Prize soon. So I vividly remember that, when I first presented my very own time-resolved data obtained in his laboratory to him, I was quite embarrassed because they were really noisy and not that convincing or conclusive. Much to my surprise, AZ responded, "Don't worry, first data are always noisy." This struck me as so true the longer I have worked as an experimental scientist.

Another encounter further demonstrated his great esteem and passion for experimental data. Finally, at one point, we got together to write a paper and my major task in this endeavor was to prepare some of the figures. I took great care to display the required information completely and concisely, but I was also very eager to be as perfect as possible with formal details such as the line widths or getting the size of the axis labels just right to be displayed properly in the final journal publishing format. In order to demonstrate that I had prepared the figure appropriately for the journal, I included it in the manuscript draft in the one column width appropriate for publication. Of course, in this way the figure only filled at most one quarter of the paper sheet. When he noticed this, AZ said, "Don't do it this way. We are proud of our data; let's show them in full letter size in the manuscript."

As a last anecdote, I remember one of the Friday evenings at which AZ would invite us after the group meeting to some beer at the Caltech Athenaeum. On one of those occasions, he asked everyone to give a brief answer to the question, "What makes a good scientist?" There were many intelligent proposals, but I think I offered a pretty dull one claiming that the proper implementation of scientific methods is one of the major requirements of being a good scientist. Therefore, I thought, I might use this occasion here to extend and reconsider my answer to the question. I now think that a good scientist is characterized by curiosity, by asking the right questions, by excitement, and most of all by a vision to picture his work in the right context. To say it in the words of Carl Sagan, "Take a step back and look at the big picture." To me, AZ always had the vision and he was well aware of the big picture, and I am grateful to him for promoting it to us.

I am very obliged to my Caltech fellows Dongping Zhong, Daniel Hern Paik, Udo Gomez, Vladimir Lobastov, Torsten Fiebig, Nam Joon Kim, Spencer Baskin, and all the rest of the Zewail group of around 1999. I am indebted to the Alexander von Humboldt Foundation for continued financial support.

Thorsten M. Bernhardt is professor of physical chemistry at the University of Ulm, Germany, since 2005. He received his Ph.D. from Humboldt University Berlin, Germany, in 1997. After spending two years as a postdoctoral fellow with Professor Ahmed Zewail at Caltech, he joined the laboratory of Professor Ludger Wöste at the Free University of Berlin and obtained his habilitation degree in experimental physics in 2006. His research focuses on the exploration of size effects in finite systems. He applies mass spectrometry and ultrafast laser spectroscopy to study chemical reactions on small metal clusters in the gas phase and at oxide surfaces.

29 A Great Mind, a Great Mentor

Steven De Feyter*

Humor... a great sense of humor! That's what I remember when I look back at my time at Caltech. It was 1997. I'm a Ph.D. student in the group of Professor Frans De Schryver at the University of Leuven (KU Leuven), Belgium. Ahmed visited Leuven as he was poised to receive an honorary doctor degree. Frans De Schryver had a knack for selecting scientists for an honorary degree. Three of them received the Nobel Prize later on in their career. Ahmed would receive the Nobel Prize two years later.

I was in the last year of my Ph.D. study and was looking for a good place to do my postdoctoral research. I didn't find the place; the place found me. It's not what it seems though. Ahmed was definitely not impressed by my femtochemistry skills — I didn't have any experience with ultrafast spectroscopy. I'm sure that my promoter had talked to Ahmed about one of his students (me) who still had a lot to learn, but that it would be a great opportunity if he would get the chance to go to Caltech. And so it happened.

March 1998 — I arrived together with my wife in Los Angeles International Airport. Thorsten, one of the postdoctoral researchers working in AZ's lab, picked us up and brought us to the Vagabond

* Postdoctoral Research Fellow.
Email: steven.defeyter@kuleuven.be

Inn on Colorado Boulevard. Colorado Boulevard didn't really fit the picture we had in mind of California. What a contrast it was with the Caltech campus.

Science — the plan was to combine ultrafast spectroscopy and scanning tunneling microscopy. Or at least, that was what I thought the plan would be prior to my arrival in Pasadena. Plans had changed though. Ahmed decided to have me working on organic femtochemistry. Having no experience with colliding-pulse mode-locked lasers, vacuum chambers, Rice–Ramsperger–Kassel–Marcus theory, and so on, I was teamed up after a few weeks with Eric Diau, a great postdoctoral scholar. We soon turned into a good team and I learned to master the dye lasers and much more. It was a very rewarding period, both scientifically as well as socially.

Things went well and after a few months I was proud that we could present our first manuscript to Ahmed. In an effort which one could call "team writing," Eric and I did our utmost best to submit the best possible manuscript we could come up with. We nailed an appointment with AZ to discuss. Great was my surprise when I learned that AZ apparently hadn't read the manuscript. Instead, we were asked to go to the blackboard and discuss the content of the manuscript. The outcome of the discussion was that we needed to redraft the manuscript, and the same story repeated itself a few times. Finally, Ahmed took manuscript version number X and started to correct it while we were in his office. He only corrected the introduction, the discussion, and the conclusions (in red, writing between the printed lines). He must have had confidence that the description of the experiments was fine.

November 1998 — it really sounded like a good idea to visit the Grand Canyon. I made plans for me and my wife to leave the day after Thanksgiving for the Grand Canyon. Now, it turned out that, at that time, we were preparing and discussing another paper. Ahmed offered to discuss the manuscript on Saturday. But what about our Grand Canyon plans? I asked Ahmed if it would be okay to postpone the discussion as I had made travel plans. He agreed… but it took several months before he wanted to restart the discussion on that paper again.

Despite my "laziness," I believe that Ahmed was quite pleased with my performance. Though I had originally planned to stay only for one year (till March 1999), he invited me to his office one day and he asked me to stay longer. This was not part of the plan. We were in the process of selling our car, hadn't renewed our rental contract of the apartment, and had already purchased our tickets to return to Belgium. My wife was also planning to resume her job again in Belgium. A few days later though, we decided to accept Ahmed's offer and stay a few months longer. But what about the expenses we already made to return to our home country? So, I decided to ask Ahmed if he was willing to cover some of those expenses. He looked at me for a few seconds, smiled, and said, "Steven, consider this as an investment in your future." Discussion closed. It turned out that Ahmed was right...

Those group meetings. Why was it necessary to schedule group meetings on Friday in the late afternoon? That doesn't work in my group now, though it worked in Femtoland. Those group meetings were often quite pleasant. He would tease the Europeans who were only interested in holidays and the Asians because they didn't take holidays.

What has affected me most as a person are the scientific discussions in his office and indeed his sense of humor. Ahmed, thank you for the chance to grow as a scientist and person in your group. Thank you for your continuous support.

30 My Journey through Zewail's Femtoland

Boyd Goodson*

I was a member of the Zewail laboratory as a postdoctoral scholar from 1999 to 2002. I am currently a professor of chemistry and adjunct professor of physics at Southern Illinois University at Carbondale.

I first met AZ when he came to Berkeley to give an invited talk while I was a Ph.D. student there. Then, I was working in the laboratory of Alex Pines, an old friend of AZ's since the 1970s. I believe that Alex was a newly minted professor at Berkeley at that time while AZ was a postdoctoral researcher in the Charles Harris group, and the Harris laboratory was just across the hall in the sub-basement of Hildebrand Hall. While I had long known of Ahmed Zewail's work (who didn't?) and kept up with my "ultrafast interests" since my undergraduate thesis days at the Warren laboratory at Princeton, I admit I was also intrigued by the personal connections that AZ had to Alex, Warren (AZ's former postdoctoral researcher), and Berkeley; in fact, the subtitle of AZ's talk was "Triggers from Berkeley." It was neat to interact with him at lunch and I very much enjoyed his talk, but at that time, I thought that was that.

Fast forward to my last few months in the Pines laboratory when I was beginning to explore options for postdoctoral study; I sent out a number of applications — "cold-calls," really — to laboratories

* Postdoctoral Research Fellow.
Email: bgoodson@chem.siu.edu

I was really interested in joining, often without even knowing if there were actually any positions open. I remembered my positive interactions with AZ during his Berkeley visit and despite the fact that I had no experimental background in femtosecond laser spectroscopy, I thought, "What the hell, why not try…?" While reading more about the ongoing projects in the group, I became particularly interested in the ultrafast electron diffraction (UED) project, and I highlighted "UED" in my application letter.

Some days later, I remember trudging into the Pines laboratory common room, only to look up and see the following words scrawled by my lab-mate Megan Spence on the blackboard: "ZEWAIL CALLED!"

Heart suddenly pounding, I grabbed the laboratory telephone and excitedly returned his call to schedule an on-site interview.

Not having ever been to Caltech before, I was alarmed to discover that the taxi had dropped me off on the wrong side of campus. Suddenly fearing that I would be late for my interview, I set off running (despite the heat, in suit and tie no less) in the general direction of Noyes Hall — only to realize just how small Caltech is; I got there so quickly that I ended up being 10 minutes early. Needless to say, the interview went well and AZ offered me a position in the UED laboratory! But now I had a decision on my hands — I had also just gotten an offer from a laboratory at Harvard in a field much closer to my graduate work. After all, what the hell did I know about electron diffraction? But the once-in-a-lifetime opportunity to work with AZ was simply too much to pass up. I threw caution to the wind and excitedly accepted AZ's offer, and thus I started at Caltech that August.

As planned, I joined the UED team — specifically working with the guys who were finishing up the construction of the new UED3 apparatus. While in many ways it was a completely different environment and laboratory culture from my graduate school days, I hit it off well with the UED team members and settled in nicely. While my thoughts were initially set on somehow trying to study the structural dynamics of proteins with this device, instead we targeted smaller, more realistic "game" — like $C_2F_4I_2$ and cyclohexadiene — and after results started rolling in, our first paper with this device was published in *Science*.

In the fall of 1999, I remember hearing about how AZ was recently named into the Pontifical Academy and there was a strange feeling in the laboratory — a nervous, excited buzz — that maybe this year really would be The Year. We quietly joked that perhaps the Pope had some "inside knowledge" from above? We knew that the announcement for the Chemistry Nobel Prize would be available online sometime around 6.00 a.m. local time on that appointed day in October. Although it might seem a little ridiculous now, a number of us thought it would be fun to wake up extra early and get the news in real time from the official Nobel web site — just in case. I remember sitting groggily in front of my computer, impatiently clicking again and again to refresh the screen in anxious anticipation. Suddenly there it was, in black and white: "Ahmed Zewail: Winner of the 1999 Nobel Prize in Chemistry." At first I didn't believe what I was seeing — what were the chances, could it be true? But there was no doubt about it. I got to the laboratory as fast as I could (as did everyone else) and we were giddily trying to figure out what to do. A bunch of us ran out for champagne and I, not knowing anything about wine, grabbed the prettiest, most expensive bottle I could find (or at least afford!): *Veuve Clicquot* (a bottle which, I'm ashamed to say, I absent-mindedly left behind after I attended the Memorial Symposium this winter). We all celebrated with AZ and he was rightfully beaming with his big, irrepressible smile. I was so excited for (and proud of) him in achieving a kind of scientific immortality, that he finally had this recognition after decades of pioneering work.

Word soon spread like wildfire and congratulations came pouring in from everywhere — and I do mean *everywhere*, all around the world (including a celebratory cake from a certain someone at Berkeley). Of course the press came from all around too and later that day there was a hastily thrown-together press conference at Caltech. Various campus dignitaries were there, including then-Caltech president (and Nobel laureate in his own right) David Baltimore. I sat in the back row with my lab-mate Vladimir Lobastov and we enjoyed taking in all of the day's events as they transpired. As the press conference went on, reporters kept pressing AZ for layman explanations of femtochemistry — how it works and what it means.

As anyone who has worked in femtosecond spectroscopy knows, this can be a bit of a struggle, particularly when asked what good this rather fundamental scientific work has accomplished for the common man in their everyday life. Faced with one such persistent questioner, AZ explained that with ultrashort laser pulses, holes can be drilled in teeth for the treatment of cavities and the heat from the laser would dissipate so quickly that the patient won't feel it. Vladi and I instantly looked sideways at each other, grinning, and I remember thinking, "Uh-oh, now the world is going to think we do femto-*dentistry*..." Later that day, many of us were interviewed by various press organizations and sure enough, when the local Fox affiliate did a five-minute television segment on AZ and his Nobel Prize (including filming the UED3 apparatus in action), they opened their broadcast with a shot of a dentist chair, saying something along the lines of "imagine going to the dentist office to get a cavity drilled and not feeling any pain!" They also ended the segment by saying that AZ's work on sodium iodide would somehow "lead to the creation of salt that tasted good but didn't have any of sodium's negative health effects." Face-palm! Oh well, the press coverage served to get the word out about femtochemistry, but it also taught me a memorable lesson that when it comes to discussing science, it can be very difficult for the press to come remotely close to getting it right, so we scientists have to be careful with our words and help them out as much as we can.

After several months, the roar of excitement of the Prize finally began to die down, AZ was around more often, and things naturally fell back into a pretty standard routine. Unsurprisingly, the UED project was of particular interest to AZ and we checked in with him regularly. First thing almost every morning, we would start the day by hitting the Red Door Café to get some coffee, and we would stop by AZ's office on the way to see if he wanted his regular order: usually a non-fat cappuccino or latte, "extra hot." Although AZ certainly relished his coffee, I think his order also had the secondary purpose of making sure we didn't linger *too* long enjoying *our* coffee in the sun at the Café. We were always careful to make sure his was still piping hot by the time we returned. [Note: while the validity of the "extra hot" order was recently questioned by other group members at the Memorial Symposium, I would have to say that I tested this hypothesis once myself by ordering my mocha "extra hot" and despite reflexively spitting out my first sip, I scalded myself so badly that the inside of my mouth felt like an old rubber boot for several days, telling me that (1) "extra hot" *did* indeed mean something, and (2) AZ was made of tougher stuff than I was!].

While the experiments were going on (which was most of the time — and I do mean *most of the time*, since UED experiments were 24-7 affairs that would drag on day after day to get enough signal), AZ would frequently pop by our laboratory unannounced to check on how things were progressing. He would ask a few questions and then urge us to just carry on (as if he weren't there somehow), and with a grin and a twinkle in his eye, he would say that he just wanted to "smell the cooking" — a phrase I now use regularly when checking in on my own group members. When it came time to writing papers, AZ worked with us directly in a very collaborative way. We would sit for countless hours in his conference room, meticulously going over painstakingly crafted figures, sometimes hammering out each line word by word — and I'm proud to say that AZ often relied on me to help craft key turns of phrase and endow each sentence with every bit of meaning that I could. We would often have good-natured arguments about what I thought we could (and could not) "get away with" saying, in terms of what conclusions we could glean from a given experiment. Admittedly, I would often err on the side of shoe-horning every detail and hedge that I could think of into a given discussion, whereas AZ preferred to keep things as simple as possible, focusing on — to unapologetically steal a phrase from *A Beautiful Mind* — the "governing dynamics" of a given underlying phenomenon. At moments of congenial conflict like these, he would often smile and refer to me as "Boyd the Lawyer" — a moniker I accepted with pride.

When I wasn't working on (or "baby-sitting") the UED apparatus or preparing a sample, I would often be analyzing data or performing quantum chemical calculations in the computer room, which was sometimes teeedious, taxing work. I was a big fan of the satirical faux-news site "The Onion" and when I took a break sometimes

I would blow off some steam and get a laugh from reading its humorous articles online. One morning, I couldn't help but click on an article entitled, "National Science Foundation: 'Science Hard'"; basically, it was a satirical article about how there was now world-wide consensus among scientists that science was just too hard and esoteric to be worth pursuing, confirming the "Science-Is-Hard Theorem." But sitting there in the Zewail laboratory, I nearly choked on my coffee when I read:

> Dr. Ahmed Zewail, a Caltech chemist whose spectroscopic studies of the transition states of chemical reactions earned him the Nobel Prize in 1999, explained in layman's terms just how hard the discipline of chemistry is, using the periodic table of the elements as a model.
>
> "Take the element of tungsten and work to memorize its place in the periodic table, its atomic symbol, its atomic number and weight, what it looks like, where it's found, and its uses to humanity, if any," Zewail said. "Now, imagine memorizing the other 100-plus elements making up the periodic table. You'd have to be, like, some kind of total brain to do that."

[Note: you can still find the article here: http://www.theonion.com/article/national-science-foundation-science-hard-1405.]

My lab-mate Jonathan Feenstra and I got a huge laugh out of that fake quote. Knowing AZ, it was (and still is) completely impossible to imagine him — in his voice — ever saying anything remotely as inane as that! Nevertheless, even today when confronted with a difficult scientific problem or setback, my mind sometimes returns to that Onion article and I'll think to myself, "Yup, Science IS hard..." (I have my original printout of this article taped to my office door with AZ's fake words highlighted.)

My three years in the Zewail laboratory went by quickly. Group members came and went at a regular pace: graduate students graduated, postdoctoral researchers left to take other positions. Caltech also had its own interviews of chemistry faculty candidates and AZ always invited us to attend their presentations. One thing that stuck with me was how AZ would always say, in reference to the evaluation of faculty candidates, that one should be a big believer in the *person* and how strong they are as a scientist and as a thinker, and be less concerned with the specific area of research they are in. Perhaps, given my lack of experience back then, this was the line of thinking AZ used when he offered me a position in his own laboratory. In any case, I found this to be an excellent piece of advice and perspective not only for how I would approach my own faculty application process, but more importantly, how I approach my candidate evaluations when *I* am on a faculty hiring committee.

In the fall of 2001, it was my turn to apply for faculty positions. While the national competition was extremely fierce (including two other AZ group members!), luckily I did get a number of interviews. But with a newborn baby in the house, it was very hard for me to travel. Although my interview at Southern Illinois University, Carbondale (SIUC) was a little strange (in ways that I still laugh about), I was very excited when they unofficially promised me a very attractive offer. To tide them over, I told them that I would cancel all of my remaining interviews except for one at a prominent Big Ten university that I had on the calendar. I enjoyed my interview at that university very much and the faculty and students there were very friendly, but as the interview wore on I got the feeling that I wouldn't be getting an offer. While in an elevator near the end of my visit, a congenial faculty host asked me what I would do if I didn't get the offer at his school. I answered that I would likely take the offer I had pending at SIUC. Somehow taken aback, he blurted out — politely, but to my face (like he was doing me a favor) — that I would be making a big mistake because if I "went to Carbondale," I wouldn't get good students, wouldn't get funding, and wouldn't be able to publish. The elevator door then opened and my host quickly escorted me to my next appointment. He then wished me well but I was so busy reeling from that psychological beat-down that I could barely pay attention to what anyone was saying. What if his warning were true? Was I about to ruin my professional life?

When I got back to Caltech, I urgently asked AZ if I could meet with him to talk about my interviews and options. He sat me down in his office and I explained that I had a strong offer from SIUC, but based on what I was hearing, I was unsure if I could be successful there. I really appreciated that AZ was able to take the time to sit down with me and talk about my job conundrum. After listening, he was careful not to try to make the choice for me, but instead asked questions and just helped me illuminate the pros and cons of every alternative. Indeed, being a believer in the *person* (it matters less where you are, but much more what science *you* do there), he did not dissuade me from the offer at SIUC and once I accepted it, he fully supported my decision. I have been at SIUC ever since and can happily proclaim that my Big Ten host was dead-wrong on every count.

I tied up as many loose ends as I could, said my goodbyes to AZ and my fellow group members, and left Caltech for SIUC that August. I never did get to try UED on proteins (besides some promising but unpublished computer-modeling work); in fact, the closest we ever got while I was there involved the ring opening of 1,3,5-cyclooctatriene (in other words — not that close to proteins!). However, I did leave with several papers co-authored. More importantly, however, I learned so much while in the Zewail laboratory — not only from AZ himself, but also from all of my fantastic lab-mates that I shared experiences with during my time there. To this day, I remain extremely thankful not only for the opportunity to work in his laboratory, but for all of the help and advice AZ gave me along the way.

However, I don't think that AZ ever knew that, in a strange way, he also helped me complete what was arguably the most important step of my life — years before we even met! Back when I was an undergraduate at Princeton, I had a girlfriend from my hometown in Greencastle, Indiana. As high school sweethearts, we maintained a long-distance relationship throughout those years. By the time I was a senior, things had gotten serious and I decided that I would ask her to marry me. But there was just one silly yet seemingly unavoidable obstacle: I had no money to get an engagement ring and no idea how to come up with it.

Fast forward to near the end of the school year: I heard that the Department of Chemistry had a number of endowed awards for graduating seniors. One of these is the McKay Prize in Physical Chemistry, with the winner (or winners) determined from the responses to an essay-based exam. Given that I was a physical chemistry student, I thought it was my best chance. When I took the exam, the question in front of me basically asked: "What do we know of transition states, and how do we know they exist?" In my answer, I filled the pages with everything I had learned from Warren (or at least everything I could rack my brain to remember) about the work of Ahmed Zewail using femtosecond lasers to probe ultrafast dynamics of molecules and their transition states — the very work that would later earn AZ the Nobel Prize — and the experiments performed, no doubt, by many of the other folks contributing to *this* work. After some anxious waiting, I found out that not only was I a co-winner of the McKay Prize, but that it came with a cash prize of ~$2000! I was ecstatic; in what seemed like only femtoseconds, my fortunes had changed and my future had become suddenly clear — I now had enough money to get a proper engagement ring. I proposed after I got home that summer (she said yes!), and we got married a year later — and of course are still together to this day, now with four children in tow. Indeed, our first son's name is a hidden tribute to both Alex and AZ: his middle name is Alexander, and his first two initials are AZ spelled backwards.

Looking back, I very much enjoyed my three years with AZ at Caltech and have many fond memories. And while it was a sombre occasion, coming back to Caltech and joining everyone to celebrate AZ at the Memorial Symposium was an amazing experience. It also served as an apt reminder of how things change during the passage of time — one of the themes of AZ's life's work. While not everyone could make it back, I was happy to see so many familiar faces, most of whom I hadn't seen since the day I left, or even earlier. I'm grateful for the wonderful group picture — what a huge family we have become! It was also neat to see the laboratory where I worked one last time, and to see that the UED3 apparatus — and all of our notebooks — were still there! Lastly, I made sure to get a photo of the "WELCOME TO FEMTOLAND" sign at the corner of the Noyes subbasement hallways, an emblem that surely belongs in the Smithsonian. The Symposium was a magnificent tribute to the man and his legacy, and I am most grateful that I was at least a small part of it.

Photo 30.1. Ahmed Zewail and Boyd Goodson, taken at a reception celebrating the 1999 Nobel Prize in Chemistry.

Photo 30.2. Polaroid of Ahmed Zewail and Boyd Goodson, taken in the Noyes sub-basement, in front of the UED3 device.

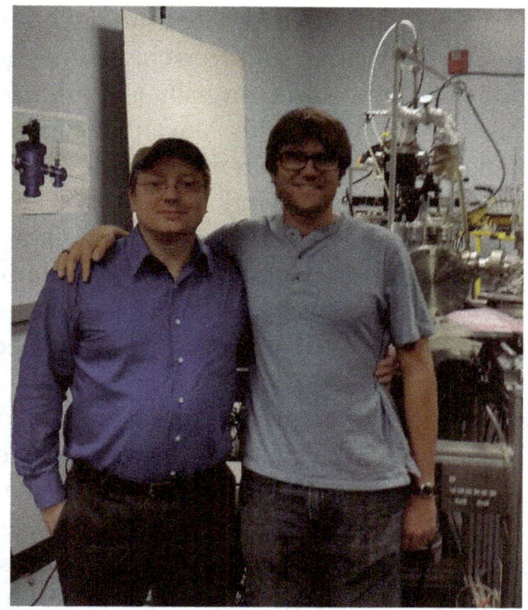

Photo 30.3. Boyd Goodson and Jonathan Feenstra in 2017 (after the Zewail Memorial Symposium), in front of the (still intact!) UED3 device.

Photo 30.4. The "WELCOME TO FEMTOLAND" sign that graces the sub-basement of Noyes Hall.

Born in Greencastle, Indiana, **Boyd M. Goodson** did his undergraduate thesis work with Warren Warren and Herschel Rabitz and graduated *magna cum laude* from Princeton University in 1995. He then earned his Ph.D. in chemistry in 1999 with Alexander Pines at University of California, Berkeley and the Lawrence Berkeley National Laboratory. Following his postdoctoral work with Nobel laureate Ahmed Zewail at Caltech in the field of ultrasfast electron diffraction, in 2002 Goodson joined the chemistry faculty at Southern Illinois University, Carbondale (SIUC), where he was later promoted to associate (2007) and recently full professor (2014); he was also named an adjunct professor of physics in 2011. At SIUC, Goodson's research interests have concerned the development and application of techniques for enhancing the detection sensitivity and information content of nuclear magnetic resonance spectroscopy and magnetic resonance imaging, with a particular emphasis on hyperpolarized contrast agents for *in vivo* molecular imaging. Goodson has co-authored over 60 scholarly articles and over 110 presentations at conferences in North America and Europe, with an external funding of nearly $5M. Recognitions of his research and teaching include a National Science Foundation CAREER Award, the Cottrell Scholar Award and Research Innovation Award from the Research Corporation for Science Advancement, and the Kaplan Research Award from the Southern Illinois Chapter of Sigma Xi. He was also the invited evening speaker for the Illinois State Science Fair and the Illinois Junior Science and Humanities Symposium in 2015 and 2017, respectively. Goodson lives in Carbondale, Illinois with his wife Amy and four children: Elysia, Zachary, Jonas, and Olivia.

31 Small Encounters, Big Impacts

Shouzhong Zou*

In one's life, many short encounters leave big impacts. Although I did not have many conversations with AZ, I benefitted tremendously from various direct and indirect interactions with him. He imparted his wisdoms and insights to us through many speeches at group meetings and shaped our philosophy through personal examples. I am most grateful for the superb research environment he provided for my career development. Because of his research group, I had many opportunities to meet top scientists in various fields and my close interactions with several group members at the time taught me a great deal about the field of ultrafast spectroscopy and life in general. Some of them have become friends to last a lifetime. It was certainly an important, transformative chapter of my life.

With a strong recommendation from my Ph.D. advisor Michael Weaver at Purdue University, I joined the Zewail group as a joint postdoctoral fellow between AZ and Fred Anson, a towering figure (in both electrochemistry as well as physique) at the Laboratory of Molecular Sciences (LMS) in early October 1999. Mike did his postdoctoral research with Anson in the early 1970s and he strongly encouraged me to seek a postdoctoral position at Caltech because it is regarded as "research heaven." With this idea implanted in my

* Postdoctoral Research Fellow.
Email: szou@american.edu

mind, I inquired about the possibility of joining his group when I met Fred at an electrochemistry conference in January 1999. I was very fortunate that Fred was looking for a new postdoctoral researcher to fill a vacant position at LMS. I was invited to visit Caltech and give a seminar at LMS. At that time, I was studying carbon monoxide (CO) adsorption on various single crystal electrode surfaces of Pt-group transition metals using surface infrared spectroscopy. During my talk, AZ asked me how CO bonds to metal surfaces. I told him that it is similar to the metal carbonyls and he continued to press on. The bonding between CO and metal ions is a classic example in organometallic chemistry and is covered in most inorganic chemistry courses. It involves the σ donation from CO to an empty metal *d* orbital and the π^* backdonation from a filled metal *d* orbital to a CO π^* orbital. I felt a little bit intimidated at the time, thinking that he was just testing my understanding of the topic. In retrospect, based on his conversation with Anson at the time, I should have realized that he was simply trying to understand the very basic yet central problem of chemical bonding. After I joined the group, there were several occasions where I observed him asking very rudimentary questions. Getting to the bottom of the problem is his style. Although he was a top scientist, he did not pretend that he knew everything and was not afraid to ask simple questions. I often use these examples to encourage my students to ask questions in meetings and seminars.

On October 2, I arrived at Pasadena after a week-long scenic and relaxing train ride from Chicago. Incidentally, there was an earthquake the second night after my arrival. What I did not know was that another earthshaking event (to me) was coming. On the second Wednesday (October 13) I was on campus, Zewail was awarded the 1999 Nobel Prize in Chemistry. In the first group meeting after the award was announced, he assured the group that he would continue doing research and encouraged us to keep working hard. He was optimistic that there could be another award from the new ultrafast electron diffraction work. Half-jokingly, he said that once he received the Nobel Prize, people assumed he knew everything. He kept very clear about who he was — a scientist. As the newest member of the group, I had the honor of attending the celebration ceremony and the

Photo 31.1. Ahmed Zewail and me at AZ's Nobel Prize celebration ceremony.

banquet, and he took pictures with each and every one of his group members, including me (see Photo 31.1). It was such an exciting moment.

With the celebrations and many interviews, AZ was much busier than before. Luckily we had already discussed about my project. I was tasked to continue studying the kinetics of association and dissociation of dioxygen from picket fence porphyrin, a heme mimic. To start, I was asked to set up laser systems in Noyes 036 that can measure transient absorption in a time scale from microsecond to femtosecond. This involved combining an "ancient" YAG pump dye laser and one of the very first Tsunami femtosecond laser systems from Spectra Physics. Later, I learned that no one else was willing to take over this project. However, for me this was a great starting point. As an electrochemist by training, I did not have prior experience in ultrafast laser spectroscopy. It was therefore a good learning experience and helped me understand the workings and guts of these lasers inside out. In addition, I learned how to use Labview to communicate with and control different parts. The downside is that the project took a long time to produce publishable results as both the old and new lasers were prone to malfunctions. AZ understood what it took to set up this laboratory and was very patient with my slow progress. I had the opportunity to work closely with Spencer and learned tremendously from him. In the end, we finally finished the project after nearly two years of work. When I finally presented a complete manuscript draft to AZ, he was very excited about the results and presented them on several prestigious occasions.

After having two advisors (Mike and Fred) with very particular writing styles, I found working with AZ on manuscripts to be more relaxing and encouraging — a thoroughly pleasant experience. I was very used to having manuscripts returned with lots of red marks and at the end 90% of the sentences had been changed. AZ was very different. As long as the statements were accurate, he would not change them. However, AZ was very particular about figures. He required a high level of clarity and aesthetics and would ask us to change fonts, colors, and line styles. Sometimes, a figure would go through nearly ten revisions before it was finalized. More importantly, AZ was very cautious on the authenticity of the presented data — he required us to show him the raw data if the results were smoothened. Many of my practices in figure preparation in my own group can be traced back to habits instilled by AZ.

AZ treated his group like his family and he cared about his students. He had an assigned parking spot outside Noyes after his Nobel award. Although many of us took his spot with the wrong assumption that he was out of town, he never complained about this. I had the honor of attending a party at his house where he and his wife Dema were great hosts. At the end of the dinner, he asked each of us what we would like to have, coffee or tea, just like a waiter in a restaurant. He then served each of us with our drinks. I still have vivid memories of the way he talked to us, paying full attention and looking right into our eyes. I certainly miss his big laughs and his pats on my shoulder for encouragement. May he rest in peace!

Shouzhong Zou is Professor and Department Chair of the Chemistry Department at American University in Washington DC. He earned his B.S. in chemistry in 1991 and completed his M.S. studies in 1994 in Xiamen University, China. He received his Ph.D. in chemistry from Purdue University in 1999 under the direction of Michael J. Weaver. He then did postdoctoral work at Caltech with Fred C. Anson and Ahmed H. Zewail. He started his independent research as an assistant professor in 2002 at Miami University (Oxford, Ohio) and was promoted to associate professor in 2008. He joined the American University in the summer of 2015. His research interests include developing catalysts for low temperature fuel cells and CO_2 reduction, advancing spectroscopic and microscopic techniques for the characterization of surfaces and interfaces.

32 Impact of a Nobel Prize on the Far Side of the Earth

Samir Kumar Pal*

The event was a telephone call from the Secretary-General of the Swedish Academy of Sciences, congratulating Professor Ahmed Zewail (our AZ) on the unshared award of the Nobel Prize in chemistry at the dawn of October 12, 1999 in California. The impact of the call in the late evening on the far side of the earth, India, where the time difference is about 12 hours ahead compared to California, was remarkable. The post on the internet was like a very short excitation pulse which still persists in my eye. I came to know of the news from my thesis advisor, Professor Kankan Bhattacharyya, senior editor of the *Journal of Physical Chemistry*, who was then a young professor at the Indian Association for the Cultivation of Science, where Professor C. V. Raman had discovered the "Raman Effect" and consequently received the 1930 Nobel Prize in physics. He rushed to our laboratory where I was trying to train an old dye-laser to produce a second harmonic pulse and said, "You know... you know... professor Zewail got the Nobel Prize this year, finally!" Although I was aware about some of AZ's works and the news was really good, I however only discovered later that I was the last person in India to register that last word, "finally." In a very short period of time, I received more than 10 emails with similar news from various

* Postdoctoral Research Fellow/Visiting Associate.
Email: skpal@bose.res.in

research institutes all over India. It struck me that the Indian scientific community was full of joy and considered the recognition of AZ as a Nobel Prize winner to be a long overdue victory. When I returned to my room that night, I came to know that was the only news our entire hostel cared about. The news changed my life, and in the following years I had opportunities to reflect on those changes following that announcement of the 1999 Nobel Prize in chemistry.

One of the decisive moments of my life was to return back to my laboratory and send an application via email to AZ for a potential postdoctoral position. This would strike most people as not the most obvious or smartest thing to do for a few reasons. First, there was the possibility that the application could have been buried under thousands of congratulatory emails from people across the world. Second, my background was in physics with a specialization in electronics, and I had only started working in the field of physical chemistry after spending a significant amount of time (about three years) in a laboratory that was largely focused on electronics and telecommunication research without being able to publish anything. Thus, the application to work in experimental biophysics was very much a shot in the dark. The third and most important reason why my application was outrageous was that I had just submitted my thesis and did not even have a Ph.D. degree yet, the standard requirement for any postdoctoral position. But I submitted the application anyway and, to my great surprise, in a few hours when this part of the world was approaching the dawn of a new day, I received a very positive reply from him. In the email, AZ said that if I could reasonably anticipate receiving the Ph.D. degree eventually, he did not have an issue with my current lack of one and he wanted me to complete the official processes before leaving to Sweden for the Prize. I was somewhat dumbfounded as I was unable to believe my good fortune. But then I recovered and quickly sent an email thanking him in a few words, when the truth was really that I had a thousand words in my mind that I couldn't convey. My department was full of joy when my advisor told everybody the news and I realized that, in my excitement, I went a full night and half a day without any sleep. I would like to start my reminiscence of AZ by quoting Rabindranath Tagore who received the 1913 Nobel Prize in literature for his literary work, *Gitanjali* (poem no. 50): "Ah, what a kingly jest was it to open thy palm to a beggar to beg! I was confused and stood undecided, and then from my wallet I slowly took out the least little grain of corn and gave it to thee." The context was that one day, the poet-beggar saw God while begging from door to door in the village path. Suddenly, God himself held out his hand to the beggar asking for alms.

The postdoctoral offer from AZ could be considered the highest achievement in my scientific career, but this recognition also added a new dimension for me — it made me proud of my birthplace, a small town 200 km away from Calcutta, where passing 10th grade with a good score and getting full meals twice a day were challenging. However, within a few hours of my arrival at Caltech after one of the longest flights I ever took, I came to know that AZ was very much familiar with my name because of its similarity to that of one of his favorite teachers. My Caltech stint starting in early 2000 was very exciting as I learned something new every day, either about the science or life under the supervision of Dr. Spencer Baskin and Professor Dongping Zhong, and close association with Dr. Jorge Peon as co-worker in room 05B of the sub-basement of Noyes Building.

My first work, the femtosecond study on a protein-DNA interaction, took an unexpectedly long time to complete. After a significant number of random walks to and from AZ's upstairs office and numerous frustrating moments in Room 05B of the sub-basement, AZ asked me a simple question during my first academic interaction with him: "In three sentences, why should the work be published?" Although this question was apparently very simple to me, I soon understood its importance when AZ was not convinced with my answer despite three attempts to come up with something. AZ was not even convinced by DNA condensation, which was a consequence of the protein-DNA interaction as explained by Amani, AZ's younger daughter who was then a young scholar in biochemistry. It took me about a month to identify a convincing answer for that "simple" question: incorporating a several meters-long DNA in a micron-sized nucleus of a cell is impossible without DNA condensation. Afterwards, life got a bit easier until Professor Bengt Norden, one of the editorial

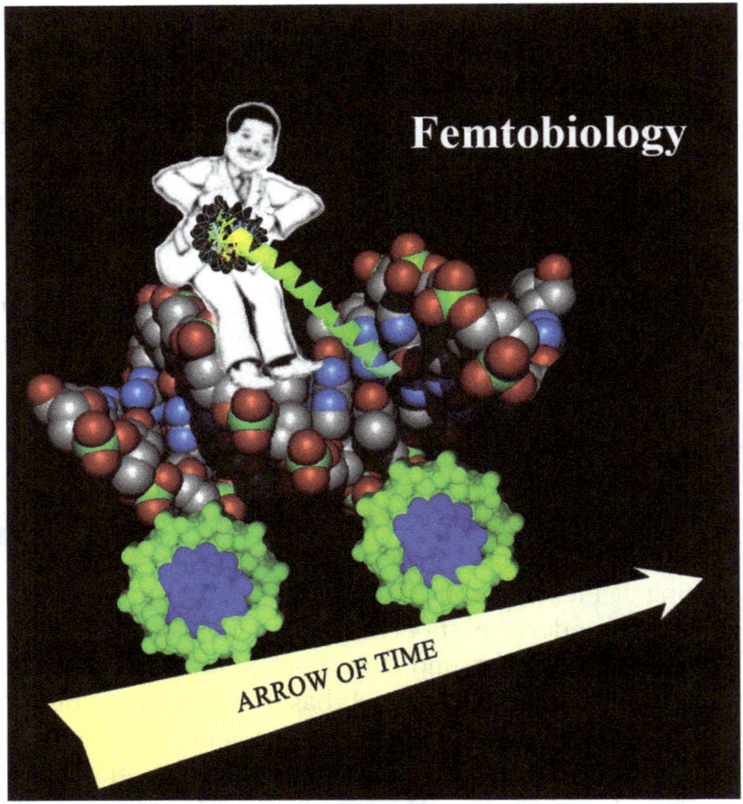

Photo 32.1. A cartoon to represent my view on femtobiology, which was inspired by a picture in our group library which was presented to AZ after his Max T. Rogers lecture at Michigan State University.

board members of the Royal Swedish Academy of Sciences, visited our laboratory. When AZ wanted me to present our work in the field of femtobiology to Professor Norden, I had a really difficult time deciding how to do it. Eventually, I came up with the idea of producing a funny cartoon as shown in Photo 32.1. The job appeared similar to praising the sun using a little candle. I started explaining our work with the cartoon and a miracle happened — I saw a big smile on AZ's face and I immediately felt immensely confident to talk in front of Professor Norden in my natural way! I tried to sound convincing about what I believed — that knowledge of ultrafast processes in biological macromolecules (charge transfer, hydration, isomerization, etc.) was important in order to control their functionality in the physiological milieu.

I was attending a class given by Professor J. T. Hynes in AZ's bigger office along with another Nobel laureate, Professor Rudolph A. Marcus from Caltech, on the mechanism of SN_2-type reactions in the liquid phase. It was not only an honor as a student to be able to sit with Nobel laureates, but I also saw AZ as real fellow student in that class. As AZ's interest in the study of femtobiology grew, the people in the world who were most experienced in the study of water started visiting our group frequently; miraculously, I had the opportunity to interact with them very closely and came to know their personal stories as well as how their discoveries happened, which was equally exciting. I was eagerly anticipating the visit of Professor Biman Bagchi from the Indian Institute of Sciences, Bangalore, when we started working on "biological water," the water molecules in the close proximity of biological macromolecules. Any attempt to express how incredible it was that these two geniuses came together and interacted is certainly beyond the scope of this reminiscence. Professor Bagchi amazed us as he taught us how he used his intuition to derive insights from a model equation that was the outcome of an involved theoretical analysis. The party merrily proceeded and eventually ended with the successful launching of the Bagchi–Zewail (B.Z.) model, which can be used to explain the lethargy of water molecules in the proximity of biological macromolecules in responding to any external excitation (see Photo 32.2).

AZ's first trip to the far side of the earth was far from simple and involved a fair bit of physical and economic cost. He had to take lots of painful vaccinations and he sponsored his own travel to India in order to provide financial aid. Following an invitation to give a talk from Professor C. N. R. Rao, a Linus Pauling Research Professor and Honorary President of the Jawaharlal Nehru Centre for Advanced Scientific Research, Bangalore, AZ decided to visit India for the first time. I suggested that he should travel through my city, Calcutta, and he agreed. Immediately, I called to inform my thesis advisor and

Photo 32.2. This picture was kindly captured by Dema, showing Professor Biman Bagchi (next to AZ) at a home party after successfully launching the B.Z. model, including Jorge and myself (left and second left) with happy smiles. In AZ's arms were Nabeel and Hani.

Professor Debasish Mukherjee, then Director of the Indian Association for the Cultivation of Science, about AZ's visit. After being briefed by Professor Mukherjee, Viren J. Shah, who was the Governor of my state, West Bengal, was delighted to consider AZ as his special guest and thereafter, a significant chunk of the issue of transportation was solved. The difficult job of announcing this visit to all the academic institutions and schools was taken care of by my thesis advisor, Professor Bhattachayya. The news of the talk became viral through local newspapers (see Photo 32.3) and I became famous once again as the first Indian to be AZ's postdoctoral researcher as shown in Photo 32.4. It was not until 15 years later, during the memorial meeting on January 19, 2017 where Dr. Rajiv Shah confirmed that he did not have Indian citizenship, before I could confirm the validity of that statement! I came to know from one of the reporters that there were 500 people inside the hall, and still hundreds were waiting outside as they were unable to get a place inside.

The Telegraph
KnowHow
MONDAY 25 NOVEMBER 2002

FAME AND FEMTO

Nobel dream

Ahmed Hassan Zewail's race didn't stop with the midnight call from Stockholm, writes Prasun Chaudhuri

Our photographer had such a tough time shooting Prof. Ahmed H. Zewail on the lectern that he managed to get hardly seven printable snapshots of the Nobel laureate. He kept complaining that he'd never shot somebody moving so fast who could beat the shutter speed of his good old Nikon. Yes, Prof. Ahmed Hassan Zewail, the 1999 Nobel laureate in chemistry, moves really fast. They say he's much faster in his laboratory in the California Institute of Technology (Caltech) where he finishes an experiment before you can bat

Continued on Page 2

Photo 32.3. One of the newspaper cuttings that highlighted AZ's first talk in India at the Indian Association for the Cultivation of Science, Calcutta, on October 16, 2002.

The non-academic side of India's warm welcome to AZ was also remarkable as the Nobel Prize had a huge impact on the minds of common folk. He received the Rajiv Gandhi memorial award in the name of our late prime minister from his wife, Mrs. Sonia Gandhi, who was then a very powerful opposition leader (Photo 32.5). After returning to Caltech, AZ shared many anecdotes of his visit to India.

In his guru's footsteps

Until recently femtochemistry appeared just an ornamental offshoot of science with no practical applications. But the new discipline is gradually breaking new grounds in all areas of science ranging from dentistry to meteorology. "Exciting frontiers are opening up in the realm of biology," says Zewail.

Perhaps the most exciting research on this new frontier is being conducted by an Indian post-doctoral fellow in Caltech called Samir Pal.

Just like his mentor, Pal was born in an obscure village in Bankura, West Bengal. Through sheer hard work he went on to be a star researcher in the Indian Association for Cultivation of Sciences (IACS). Working in the lab of theoretical chemist Prof. Kankan Bhattacharya, Pal made his foray into the femtoworld. "Much of the research he's pursuing now in Caltech began in my lab," says Prof. Bhattacharya. In fact, it was not until Pal stepped into Caltech's Laboratory for Molecular Sciences (LMS) that there was any real breakthrough of femtochemistry in life science. "Samir's works on how water molecules attach to proteins will go a long way in unravelling the universal problem of protein folding," said Zewail in his talk at the IACS.

Right now, protein folding is the biggest enigma of 21st century science. If everything goes well, Pal's insights may help invent new drugs, improved surgery and smarter scanning techniques for critical diseases like cancer and AIDS.

Samir Pal (left) and Prof. Zewail

Curiously enough, despite making forays into cutting-edge research Pal wishes to come back to his alma mater to complete the venture. "Given the opportunity, I'll show that it's possible to do top-class research in India."

Photo 32.4. Another newspaper cutting from the *Telegraph*, November 5, 2002, which highlighted my association with AZ as his first Indian postdoctoral student. In India, a "guru" refers to a religious mentor.

Photo 32.5. AZ receiving the Rajiv Gandhi Memorial Award form Mrs. Sonia Gandhi. Professor C. N. R. Rao was also present in the dais (left). (Photo provided to the author by Ahmed Zewail.)

Photo 32.6. Our president, Dr. A. P. J. Abdul Kalam, shared many ideas with AZ during the visit to India. (Photo provided to the author by Ahmed Zewail.)

He recounted, for instance, that Dema liked the shawl given by Mrs. Gandhi very much. He also retold the story of how our president, Dr. A. P. J. Abdul Kalam, wore a traditional Indian Chappal during the formal discussion with him, which did not include formal shoes (AZ and Dr. Abdul Kalam are pictured in Photo 32.6, although not during the time he wore the traditional Indian garb). I can still recall AZ's detailed description of his experience of seeing the Taj Mahal bathed in the light of the full moon.

The news of his departure forever to the far side of the world had a deep impact on everybody. On August 2, 2016, the whole of academia in India was celebrating the birthday of Sir Prafulla Chandra Rây, an Indian chemist, educator, entrepreneur, and the owner of the first ever Chemical Landmark Plaque from the Royal Society of Chemistry outside Europe. The news came to me immediately after my talk in the early morning of 3 August at the Society. An immense silence passed through the whole auditorium following the news and I felt as if I lost my father for the second time. Whatever small achievements I have today, I owe to AZ. I would like to end this reminiscence with reference to the poet-beggar's realization as revealed in the *Gitanjali*-50 by Tagore. At the end of the day, the poet-beggar returned home and found a little grain of gold in the heap of alms. The exact statement from Tagore reads, "I bitterly wept and wished that I had had the heart to give thee my all."

After obtaining a Ph.D. in physics from the Indian Association of Cultivation of Science in 2000, **Dr. S. K. Pal** moved to Caltech to work with Professor Ahmed H. Zewail, Nobel laureate in chemistry, 1999, until the end of 2003. Currently, Dr. S. K. Pal is a full professor at the S. N. Bose National Centre for Basic Sciences, India. He has received the UK-India Education Research Inititative (UKIERI), 2006 in nanosciences and the Professor P K Bose Memorial Award, 2016 from the Indian Chemical Society.

One of his research interests is focused on the ultrafast spectroscopy of molecules and materials for potential applications in environments, energy, and health. He has co-authored over 200 publications and 5 books, given more than 200 invited presentations, and has over 8700 citations with a h-index of 48 and i10 index of about 157. He is also a co-inventor with more than 20 patents. To date, 20 Ph.D. students have completed their degree and 15 are enrolled under his sole supervision. He is an editor for *E. P. J. Techniques and Instrumentation* (Springer, London) and *Advances in Physical Chemistry* (Hindawi, USA). He has served as a visiting professor for several places including Caltech (United States), TU Brunswig (Germany), University Aarhus (Denmark), Durham University (United Kingdom), and University Leiden (Netherlands).

33 Looking Beyond the Surface, My Recollection of the Man and the Work at Caltech

Chong-Yu Ruan*

I worked in Zewail's group from 2000 to 2004. It was a time of transition as Ahmed just got the Nobel Prize months before I arrived at Pasadena. There was a sense of tremendous pride in the group about the recognition for his contributions to the study of femtochemistry. At the same time, a visible shift of the group's focus from ultrafast spectroscopy to ultrafast diffraction had already begun. My background has been in atomic, molecular, and optical (AMO) physics and diffraction. After getting a degree from the AMO group in Texas, I was eager to set foot in uncharted waters. The work of table-top ultrafast X-rays and electron diffraction had fascinated me greatly. There was a prevailing myth at that time about the difficulties associated with such work as only very few groups in the world had come close to achieving results, especially the initial efforts that had been geared towards gas phase diffraction. Diffusive gas jets produced only very weak signals, even for studying the ground state. However, the notion of being able to catch the chemical reactions as they occur was a dream of Ahmed's and, as usual, once he set his mind to something he believed in and wanted to achieve, there would be no turning back.

* Postdoctoral Research Fellow.
Email: ruan@pa.msu.edu

Ahmed and his team had been spearheading the development of ultrafast electron diffraction (UED) for many years and new territories were about to be chartered with the upgraded UED system (the so-called UED III). I thus considered myself lucky to join the UED group at the right time. Getting the undivided attention of a Nobel laureate in the midst of these hectic new developments was certainly no small feat, but I managed to develop a close working relationship with Ahmed over time. However, this relationship did not take off to a good start. I can recall the day when I first spoke with Ahmed, which was in early February 2000 during my last semester at Austin. When his secretary called to inform me that Ahmed was on the phone and would like to speak with me about my application, I requested to defer it as I was about to give a group seminar. I had been preparing for that talk all morning and it didn't immediately occur to me that a Nobel laureate probably doesn't usually have to wait when he asks to speak to a Ph.D. student. Upon hearing my request, the secretary sounded a bit surprised and repeated where she was calling from. To cut a long story short, I finished my presentation first and then told my advisor that I might need to excuse myself early from the meeting to return the call. He was very much surprised about what I did and urged me to leave and return the phone call. It was the first time I heard the calm and clear voice from Ahmed and the world suddenly felt a bit different when I received the words that he wanted to offer me the research associate position in his group at Caltech.

When I finally arrived in Pasadena in May, the purple Jacaranda trees were blooming on campus. I had a strange feeling about the place as I drove through Arizona and arrived at Caltech near dusk, which seemed to carry with it the dry warmth of the desert. The transition from Joshua trees and cacti to an unusual combination of tall palm trees surrounding the Spanish arcade-style buildings with the still-wintry San Gabriel Mountains in the backdrop gave the Caltech campus an uncanny appearance of forwardness. The Caltech campus was like a beautifully manicured garden and the place was full of energy. When I came to the Division building the next morning, I was initially apprehensive to knock on Ahmed's door as I had seen him pacing across the hallway from afar and he might have already noticed my arrival. He was talking to his secretaries on both sides of the aisle at the time as if there were urgent matters going on. I only realized later that this was very much the same every day when he was in town. The Nobel affairs never really ended until a year after the Prize — Ahmed was still receiving plenty of invitations and travelling so much that, after my first meeting with him, a long time would pass before we could really speak again. When he had more time to seriously discuss my project with me, that's where our real conversations began. Ahmed was obviously testing the waters with me as he got me to talk about where I believed the field should be 10 years from now as well as possible new directions the field can take. I proposed a few ideas that I had already pondered for awhile. By the time I reached my final year of graduate work, I had read a substantial number of papers on ultrafast electron and X-ray diffraction. Femtosecond laser technologies were maturing and the two leading groups racing to make the first molecular movies were Kent Wilson's group in San Diego, which was leaning toward the X-ray approach, and Ahmed Zewail's group at Caltech, which focused on the electron approach. I got an interview invitation from the Wilson group, but the group leader's health had suddenly worsened; it was clearly not a good time to visit.

Because of my early fascination with both techniques, I had pondered a fair bit about how similar both approaches were in terms of their applications, even though I hadn't personally experienced using either of them yet. My expectations at the time were probably unrealistic as I was especially fascinated with treating electrons as an excitation source, such as in X-ray and photon-based experiments, due to their obviously high instantaneous fields as the electrons traversed through matter. Similar to the ultrafast spectroscopy work from Zewail's group that I had been reading about, could it be feasible to drive the chemical reactions through their impulsive electric fields? Indeed, in conventional electron spectroscopy, the swift electron had been used both as the pump and as the probe. When I finally got the chance to speak with Ahmed in person, I told him about my electron pump idea and the possibility of using the dynamical wave grating to influence the molecular processes that may be observable in the ultrafast electron diffraction experiments. I did not realize how energetically incoherent the photoelectron pulses generally were, but Ahmed nevertheless encouraged me to formulate my ideas more. After scribbling down my ideas and doing a few quick calculations on paper the next day, he appeared pleased and scheduled a meeting to discuss these ideas further. The discussion went from a state of excitement to being trapped in details rather quickly. I realized that while I was trying very hard to be precise about what I meant in my formulation, he was thinking broadly and had already began to view the results in a different light from what I originally had in mind. I might have interrupted him a few times as I expressed my reservations about going too far ahead with those new directions, but he insisted that I should listen more carefully to what he had in mind. He told me a story about how the heterodyne detection technique, when applied in ultrafast spectroscopy, led to some real magic. I was too young at that time to read between the lines; only much later did I realize that his analogous comparison between the reference wave grating and heterodyne technology was correct. In the end, the conversation did not go so well. After what must have been an hour-long struggle, I realized that I had problems communicating my ideas to him. He was clearly frustrated as well. I went back down to my office feeling rather discouraged after that very first one-on-one discussion with him. But near the end of the day, he came down to my office, put his hand on my shoulder, and told me that we would try discussing it again later. Indeed, we did, and from then on we had many more great discussions, but the same topic never came up again until two years later when we were planning a new UED system.

For many years, I wondered why the communication between me and Ahmed eventually improved. As much as I want to believe it was my effort in trying to align myself with his thinking style, I realized later that Ahmed played a significant role in improving the circumstances by being a really good listener. At that time, the Zewail group had about 15 postdoctoral researchers and students working on several different projects. During joint discussions, I noticed that he would speak to one of us with one explanation and reiterate it to another with a different explanation. He was a good explainer on so many levels — his familiarity with the relevant topics aside, he was also naturally curious about how different people think and reason. He reached out to me on more than one occasion about how to communicate more effectively. In particular, he noticed that I would sometimes talk about my conclusions without explaining enough.

During the later years of my time at Caltech, especially during the time when we were thinking about building a new instrument to work on large molecules more easily, Ahmed and I started to have early morning coffee meetings whenever he was in town. One problem that he was perpetually interested in was how energy redistributes in a complex network and how different types of bonds would affect the dynamics. He had spoken on many occasions about his desire to watch protein (un)folding in real time using ultrafast diffraction techniques. Of course, this would be very difficult to achieve with the ultrafast electron diffraction machine we had back then. In such a system, the electron beam scatters from the diffusive molecular jet in incoherent superposition, and much of the structural information would be lost in the integration of signals. While we had many successes in probing the dynamics of smaller molecules, introducing larger molecules adequately into a gas jet without aggregation was clearly a formidable issue.

After working for almost two years on the gas phase UED project, I wanted to wrap things up and return to Taiwan. While preparing to apply to a few faculty positions there, I asked Ahmed for a recommendation letter. However, he told me that I was unlikely to succeed without the resources at Caltech. I knew that he probably meant well and I might have also underestimated the funds I had to raise to construct a new UED system. He persuaded me to stay for three more years to develop the new UED system with improved sensitivity. The ideas that were floating around included, first, applying the solid substrate as a template to enhance the signatures of molecular diffraction from the systems of interests as adsorbates on surfaces and, second, creating a sharp, hence more localized photoemission gun, but using the laser rather than the electric field to gate the emissions of electrons from a tip. After spending two years on the UED project, I had become more cautious about how a simple idea might turn into a massive project, but the thought of having his unreserved support to start a novel project in an uncharted territory was instantly appealing. I agreed to stay two more years, during which I would do the best I could to help realize at least some of the ideas we had talked about, and then leave for a more permanent position elsewhere afterwards.

We decided to pursue the first idea first. I was teamed up with two excellent postdoctoral researchers, Vladimir and Franco, and one very smart graduate student, Sonya, to develop the new system now known as ultrafast electron crystallography (UEC). It was obvious from the design of the sample goniometer that we were shifting from the gas phase to solids, although the real drive behind the project was to provide a platform to make the study of massive molecular frameworks from water, adsorbates, to macromolecules easier. The existence of an interface provided a means to engage different excitations, and this included not just the direct photoexcitation of the molecules but also the heat jump from the surfaces — an emerging technique to initiate protein (un)folding at the time. We started consulting experts on campus about electron diffraction at grazing incidence and the preparation of molecules on surfaces, including the introduction of waters *in situ*, even just temporarily. There was a chance that we were driven by pure adrenaline (or blind courage) as none of us were particularly experts in these areas, but after relying on our limited knowledge to affirm some of the general feasibilities, we commenced the project with the aim of putting together all the necessary components first. We started the design of a low temperature goniometer, UHV system with load-lock sample transfer, a molecular doser, and a surface preparation and cleaning stage from the very beginning of the project. We decided to retain the design of the electron gun and the camera so that, in the event that the new plan did not immediately work out, we could still use the chamber to conduct UED experiments on film without complications. There was a tremendous joy in learning new things every day, and the system started to come together a year and a half after the initial design. However, the unforeseen difficuilties in getting everything working together for such a sophisticated instrument and attempting uncharted modes of experimentation led to the most trying times in my career. We were struggling to realize our ambitious goals. The intense and varying diffraction patterns of the different substrates from which we were to obtain exclusive surface sensitivities were difficult to control precisely. Only after many more years in the field did I later learn that, to realize the heterodyne idea we had conceived of originally, the beam had to be much more coherent and a smaller-sized beam would have made control of the surface and molecular interference more manageable. But we were rather innocent about the topics at the time and were struggling for nearly a year without getting the results we hoped for.

Through this difficult time, and rather unexpectedly, Ahmed reached out to us and gave his support. On a regular basis, he would call to chat about our progresses in the laboratory and remind us that we had his support no matter what. The project eventually did work out even if not as fully as we had intended, but we did study water on surfaces and tried the various molecular assemblies before I moved on to take up a position at Michigan State University and start my own research group. The UEC system would go on to be refined with several rounds of improvements by many other talented postdoctoral researchers and students, producing many nice results over the years. It is sad to hear that the UEC system will be decommissioned at this time of writing. Looking back, things were a bit surreal, but I tend to think that what we did was not only the building of a unique machine but also the exercising of a discovery-based research methodology that was at the core of Ahmed's scientific career. In speaking with many Zewail group alumni, I frequently came across this spirit of discovery-based research methodology in their own work. I have often wondered what underlies Ahmed's confidence to pursue objectives that might not be immediately clear to many; what drives his conviction that eventually things would take shape and leap forward.

It was late January of 2004 at his San Marino house, two days before I would give my first job talk. He offered to take a look at the materials I had prepared for my presentation, but I think his intention was mostly to calm my nerves. We were sitting at his pool and our conversation topics quickly turned to those about the future. I was careful to ask him about where he saw the UED and UEC projects going as it wouldn't be fair of me to take some of our shared ideas and pursue them independently elsewhere. I had already thought about moving towards hard materials and condensed matter physics, which was how I crafted my job applications. He was supportive, but what I remembered most from that conversation was something unexpectedly different. He learned that I was fond of sketching as we looked at the trees and bushes behind his fence. He asked me how I would have started if I were to draw the trees right there and then. I replied that I would have started with the branches first. He looked at me a bit surprised and asked why, as the branches were mostly invisible from our perspective. I told him that if I were to draw the leaves first, I would not be able to put the entire tree together; establishing the imaginary branches first helps guide how the leaves of the tree would eventually shape up. He smiled and agreed with this logic of drawing. On a later occasion, he returned to this idea when he described the many burgeoning areas of science and technology. He said that they were very much like the different trees and canopies in a forest. To not lose one's grip on the important issues, one needs to look deeper beneath the surface, much like drawing trees.

When I learned the sad news of his passing, the first thing that emerged in my recollection of him was that smile I saw in his house. It very much answered my question of why he had always been comfortable pursuing what many would consider impossible during his entire career. If one could plan everything, what would be left of the meaning of discovery? — Ahmed would frequently ask, rhetorically. Of course, it takes real skill and effort to actually work through the challenges and solve the problems faced along the way. But besides all that planning and effort, most people also typically need more assurance than can be afforded by their rational instincts or burning curiosity before plunging head-on into the unknown. For him, perhaps, it is in seeing the landscapes beyond the surface and realizing a wondrous painting that started merely as a simple vision. Through him, I see a connection between my affection of art and the work of science, which always makes an impossible day in the laboratory a little more humane.

Chong-Yu Ruan is a professor in the Department of Physics and Astronomy at Michigan State University. He got his Ph.D. degree in physics from the University of Texas at Austin in 2000. Later that year, he joined the Zewail group and subsequently contributed to the projects on gas phase ultrafast electron diffraction and ultrafast electron crystallography between 2000 and 2004. He extended the methodological techniques developed at Caltech by setting up the ultrafast electron nanocrystallography system at Michigan State University. For related work, he was the recipient of the Sidhu award and the Young Researcher Award given by the International Organization of Chinese Physicists and Astronomers. He is currently merging adaptive accelerator optics and electron microscopy, which is aimed at developing table-top ultrafast electron microscope and angle-resolved electron energy loss spectroscopy systems with high-brightness photoelectron sources.

34 Ahmed H. Zewail: A Charismatic Scientific Giant

Niels Engholm Henriksen*

While doing my postdoctoral research (with E. J. Heller) between 1987 and 1988, I learned about the first seminal paper from Caltech on the dissociation of ICN, which constituted the birth of femtochemistry. Later on, after returning to Denmark, I was the co-organizer of the 1996 European Conference on Dynamics of Molecular Collisions (MOLEC). Ahmed Zewail was invited and delivered a fantastic talk. As a young professor, I still remember that I was filled with pride when he asked for a copy of one of my slides.

I was a visitor in the Zewail group at Caltech in the summer of 2000, and I met Ahmed several times over the years at the Femto Conferences (starting in the early 1990s), as well as during his visits to Denmark. In the summer of 2013, when the 11th edition of the Femto Conferences took place in Copenhagen, Denmark, Ahmed was of course again invited and arrangements were made for him to deliver the keynote lecture. Unfortunately, illness forced him to cancel.

The last time I met Ahmed was in the summer of 2015 at the 12th Femto Conference in Hamburg, Germany. He was in great shape again. During that week, I had the pleasure of participating in several dinners where he was the guest of honor. On several occasions, he was accompanied by his charming wife. The meeting of the

*Visiting Associate.
Email: neh@kemi.dtu.dk

international advisory committee of the Femto Conferences took place on a wonderful summer evening at a restaurant in Hamburg. It turned out that one of the waiters was of Egyptian descent and when he recognized the famous Egyptian at his restaurant, he became very excited. It was clear that Ahmed was a big name in Egypt and that he enjoyed his fame. Ahmed was still very keen to contribute to the development of this conference series and to that end he made various suggestions, for instance to include new members in the advisory committee.

Ahmed was obviously a giant in science. He had the ability to take scientific problems of central importance and provide "simple" explanations containing the essence of the solution. After receiving the Nobel Prize, he continued at a breathtaking speed in new directions and made outstanding novel contributions to science. I remember him for delivering fantastic talks that benefited from his great charisma, and for being very kind to colleagues including young researchers. As a person, Ahmed had an amazing personality which was unparalleled and made a lasting impression on everyone who met him.

Niels E. Henriksen (NEH) holds a Ph.D. in chemical physics from the Technical University of Denmark and a D.Sc. from the University of Copenhagen. After his postdoctoral work in the United States with E. J. Heller, he became a senior research scholar at the University of Copenhagen. Since 1991, NEH has been affiliated with the Technical University of Denmark. His research interests cover various aspects of theoretical molecular reaction dynamics including femtochemistry. In 2013, he was the chair of the 11th Femto Conference in Copenhagen.

35 For the Zewail Memoir

Jon Feenstra*

This story is probably more about me than it is about AZ, but I often tell it as an example of our relationship and of AZ's sense of humor. He often teased me as the free-spirited, laid-back American and I often served as the brunt of his jokes during group meetings whenever there needed to be a counter-example to some great tale of scientific success. This was all in jest, of course, and AZ was also fond of my good sense of humor. We butted heads a few times over scientific matters and I had a habit of not seeing the big picture quite the same way he did. AZ had a way of telling me in his academic-fatherly fashion how his thirty-something years in the field had taught him a thing or two and that my three years had not.

"This ultrafast electron diffraction work on the structure of the acetophenone triplet state is SO BIG, it deserves to be in *Science* magazine, not in the appendix of *The New Jersey Journal of Jon's Work*."

He was right and I would leave for the Red Door Café to buy him an extra hot cappuccino, as the paper indeed got sent on to be published in *Science*.

I entered Caltech and his group with the intention of carving out a career in academia, but in the midst of learning whatever I could

* Graduate Research Assistant.
Email: feenstra@alumni.caltech.edu

about chemistry and physics, I also learned that my path would eventually lead a different way. I never found out exactly what he thought about me leaving the field, and after graduating in 2006, I saw him only a precious few times. He knew me well, though; we were similar in our determination and shared, above all, an immense and indefatigable enjoyment of life.

June 2003

I was a bird-watcher throughout graduate school, as I was before and am a professional ornithologist now. As such, while working in the laboratory I would occasionally get a phone call or email about some rare bird in the area and would hope to find a spare moment to escape and see it. One day, I received a call that a Northern Parula had been found in the canyon above the Jet Propulsion Laboratory (JPL). I made sure that there wasn't anything too important on my plate and ducked out of the laboratory at about 5.00 p.m. to make a quick trip over there. I parked, hurriedly exited the car, and with my binoculars slung loosely around my neck, jogged across the JPL parking lot passing staff leaving for the day and got onto the fire road that leads up into the canyon.

Perhaps 15 minutes later while I was quietly watching and listening for this lost little bird, half a dozen uniformed NASA security guards came by truck and foot up the track.

"Freeze! Hands over your head!"

One of them relayed over radio to headquarters in a stern voice: "This is Officer —. Yeah, we got him."

I was surrounded by armed security and got ordered to stand there, spread my legs apart, and put my hands on the back of my head. One of the guards took everything out of my pockets, spread them all on the hood of the police truck, and asked me questions about my identity and purpose in the canyon. I kept asking them what this was all about and telling them that I was a Caltech graduate student out bird-watching and not whoever they thought I was. They went through my stuff, conferred with someone on the other end of the radio, kept me surrounded while passing hikers and mountain bikers gawked at me, and after about 20 minutes said that I was free to go. I was told that someone inside JPL had called security after seeing "a wild, bearded guy in a red shirt with binoculars dash into the forest from the JPL parking lot." That was me alright, trying to find my bird and get back to the laboratory. The security guards apologized for the inconvenience, returned my things, and left.

My time was up, it was going to be too dark soon, and I had to get back... without seeing the Northern Parula. The next day back at Caltech, I was meeting with AZ about something and he asked me where I had been the previous afternoon when he called the laboratory and didn't get a response. I told him the tale. He laughed as loud as I've ever heard him laugh. While laughing he asked, "You didn't tell them my name, did you?" Then he told me how his suspicious looks recently led to airport security taking away his tiny moustache scissors.

During our next regular Friday evening group meeting, in the midst of what was probably some typical heavy scientific discourse, AZ called for a break and said to the group, "and now Jon, tell us that story about that thing that happened to you this week."

36 The Beach?

Theis I. Sølling*

I still vividly remember my first day at Caltech; I stepped right into a group meeting in the room across the hall from AZ's office. It was October 2000 and I was to commence what would end up being almost two years as a postdoctoral researcher with AZ. My wife and I had enjoyed the day before on Angeles Crest Highway, amazed by our new reality in the heart of pop culture. We had gotten a bit of a tan, so when I took a seat across the table from my new boss, the world-famous Dr. Zewail, his first words to me were: "so Theis, it seems that you have been to the beach already." That was a typical disarming AZ remark. I don't think Ahmed actually believed that we would goof off to Venice but it was his way of indicating that he expected us to exhibit the same work ethic that he subscribed to himself. He showed up in the laboratory at all times of the day and would always be fully engaged in the experiments and directions of the 047 dungeons. Now, in my own faculty position I admire even more his overview of and participation in all the various endeavors of the members in his group, which exceeded 30 at that time. This was regardless of his sometimes intense travel schedule which we couldn't keep track of at all. Regardless, we did stay focused on femtochemistry-related problems and did not wander off to the beach. Ahmed always wanted to stay on top of things irrespective of time zones and country borders.

* Postdoctoral Research Fellow.
Email: theis@chem.ku.dk

When the phone rang, you wouldn't be surprised to hear AZ on the other end enquiring about matters pertaining to the merits of highly excited Rydberg states in ketones, even if it was in the middle of the night from a hotel room in Paris or in the wee hours of the morning from DC. We were always there and so was he — sand and ocean breeze isn't that appealing anyway.

There are many great memories on so many different levels. I was at Caltech when my wife and I had our first son, Andreas. I remember being a bit afraid when I had to tell AZ about it back when I first met him at a meeting in Denmark, but he truly embraced a new member of the "femto family" as he put it. His door was always open on my returns to Pasadena and I used to tug Andreas along to have coffee in AZ's office. The last time was in December 2015. AZ was as mentally fit as ever and encouraged Andreas to work hard during his upcoming freshman year in Pasadena where Janet Davis and her husband opened their home to him. It really felt like family back then, and it still does today. There are so many people who remain an integral part of our lives even after the 15 years that have passed since I left Caltech. Looking back, it has been rather amazing getting to know friends from every corner of the world and having extended stays in places as exotic as Kolkata, Hsinchu, Seoul, and even the Ruhr district.

I know that AZ enjoyed the femto family relations too and my personal story dates back to one of AZ's visits to Scandinavia. It was not too long after my return to Denmark and we had just moved back in to our apartment in the central part of Copenhagen. I remember that I was in my bedroom when my phone rang — AZ announced that he was in Sweden! I was pretty excited to hear his voice because I hadn't heard from him since I left Caltech. Ahmed wasn't too excited about my departure, something which I was of course very honoured to know, but for family reasons we had to move back to Denmark. He called from the Grand Hotel in Lund and the occasion was that he was going to receive an honorary doctorate (just one of the many he received). Before I left Caltech, we had conducted a series of exciting experiments on aliphatic amines and I had drafted a manuscript. In true AZ style, he wanted to focus hard on finishing the manuscript over the next few days. He had some business to attend to in Copenhagen and had three specific things on the agenda: first, he wanted to invite me for lunch at the Grand Hotel; second, he wanted to discuss the manuscript; and third, he wanted a lift across the bridge to Copenhagen. There are no free lunches in this world as he used to put it, so as always, AZ came to me with a clear plan. I was at the Risø National Laboratory at the time as a project researcher — a Danish euphemism for postdoctoral researcher — and was granted permission to borrow the department car to go across the sound to meet AZ the next day. The car was an old ramshackle Ford; it had once been white but was now more off-white than anything else and the interior was totally worn out. I recall feeling a little embarrassed that I was picking him up in an old wreck, but it was all that I had and it would have to do. But my worries were unwarranted; at the end of the day, AZ wasn't the kind of person who paid too much attention to such things. He was a person who always remembered his upbringing in a humble suburb of Alexandria.

AZ met me at the restaurant with that disarming, broad smile of his. There was something we certainly did share, namely a craving for Norwegian Salmon, and he had already placed his order. We then proceeded to have an excellent few hours together. It turned out that AZ's business in Copenhagen wasn't on that particular day at all, so the details pertaining to the lift were a bit murky. Ahmed had already put his comments into the manuscript and now, there was no time to waste — we were to get the manuscript done in the next few days before he was to be picked up in Lund again. I turned the car around and drove back to Denmark to immediately get hold of Carsten Kötting, my former colleague from the subbasement at Caltech. We fortunately knew what it took to complete an AZ paper. Carsten did the graphics and I did the (major) text revision. There it was — a new version was ready.

I then acquired the department car again and drove across the bridge to meet with AZ, who was waiting for me in the lobby of the Grand, and I gladly handed him the revised manuscript. I was a bit concerned this time about whether the condition of the Ford would enable us to make it across the bridge, but we turned the car around and did indeed make it to the Kingdom of Denmark. I remember pointing out our local beach so that he could rest assured back home that there wouldn't be any concerns about students wandering off to have a swim. AZ was bound for the Royal Academy and I dropped him off at the D'Angleterre, a posh hotel at the very heart of Copenhagen which was also some kind of a national pride. To ensure that I would get a little more airtime from him, I invited him for lunch in Nyhavn (New Harbour) the next day. I knew of a sneaky little place where it is, even up until this day with all the smoking restrictions in Denmark, possible to have a cigar. This was to prepare for one of his favorite rituals — to finish papers with a big cigar with all the parties involved.

The Royal Academy meeting was over and we ventured off to the cigar parlor. This time, the place was within walking distance so the old Ford was off the hook. AZ immediately brought up the manuscript and, much to my joy, it had very minor edits. It was time to fire up the cigars as another manuscript writing journey had come to an end. It was a cosy afternoon in true Danish "hygge" fashion.

AZ confessed that he hadn't called the Egyptian embassy on this trip because it would trigger a bunch of dinner invitations and, as he put it, he would rather spend time finishing the manuscript and have a cigar with me. One always felt like the center of attention in his presence. I have never met anyone with quite his people skills. That manifested itself very clearly when we took a stroll towards the car and got into a conversation with an elderly American couple who were giving the D'Angleterre's "state of disrepair" a hard time. AZ became clearly upset and told me after they had left that they had no right to criticize something which he knew to be a symbol of Danish pride. It certainly showed that he was protective of my culture and nationalistic on my behalf. After this little intermezzo, we arrived at the car and headed for my apartment. We made a visit to my wife's craftshop right next to where we lived. I remember that AZ picked up a bag of tea with little heart-shaped pieces of sugar for Dema. We rounded off the manuscript with some final edits and got it ready for submission. It had been an intense and productive couple of days and I certainly got a kick out of AZ's visit and was now fully recharged. After dropping AZ off at the airport for his Cairo-bound flight, I drove up along the ocean road, watching the waves roll ashore on the beach thinking of AZ, Caltech, and the paper we just wrote. It struck me hard how much I missed Pasadena. It was one of those moments where one doesn't realize how good it is until it's gone. Even the beaches were better compared to Denmark although we did not visit them much. Blessed be the memory of AZ.

Photo 36.1. Running the LA marathon, proudly sponsored by Ahmed with one of this t-shirts.

37 A Beautiful Bridge — Remembering AZ

Tianbing Xia*

Dr. Ahmed Zewail, or AZ as we used to call him, was and continues to be a lot of things to the world — scientific or humanitarian. For me personally, he was a bridge, and a beautiful one, too.

It was the summer of 1999, I had just completed my Ph.D. training in biophysical chemistry from the University of Rochester, and was getting ready to leave the city that is covered in snow for much of the year and head off to the sunny west coast for postdoctoral training. I had worked with the charismatic Professor Douglas Turner on RNA structure prediction based on thermodynamics of the formation of secondary conformational motifs — an approach that had always been my personal favorite while growing up talking to a theoretical physicist father over dinner every day. That is, to predict things based only on First Principles — no matter how much or how little we think we understand the nature. Stubbornness is allowed and also needed here.

I had so much fun roaming the RNA world in the basement of Hutchison Hall on the University of Rochester's River Campus, collecting tons of experimental data on the energetics of RNA secondary structural formation that would feed into the prediction algorithms that my smarter colleagues programmed. However, I would never forget one of the questions that I was asked during

* Postdoctoral Research Fellow.
Email: xia.tianbing@gmail.com

my Ph.D. qualification exam in my second year of graduate school. After hearing of my labor intensive data collection efforts, Professor Eric Kool, who had been doing "Kool" research and later moved to Stanford University, asked me, "Are you familiar with combinatorial chemistry?" After admitting that I wasn't, I probably made a subconscious decision right there and then that combinatorial chemistry would be the next thing I would learn after I got my Ph.D. with Doug — as long as the projects were also RNA-centric somehow.

This would not be a totally random decision as part of my life philosophy has always been to take on an (almost) entirely new area of science every time I move to a new place, including the move to the U.S. from China in 1994 where, on my long flight to Rochester, New York, I decided to pursue a more biologically oriented chemical field. My decision to study combinatorial chemistry involving RNA after getting my Ph.D. was an easy one, and I thought I would like to pursue a career along that trajectory afterwards. Little did I know that I would later venture into a world of sciences totally unknown to me at that time, to work with one of the greatest minds of this era of modern chemistry. Needless to say, the next major phase of my professional life was largely shaped by an unplanned journey, and a short one no doubt, with AZ. One has to realize that any journey you're taken on by a mind as great as AZ's will always be one that is too short, no matter how many years you've stayed close by his side.

The destination for my postdoctoral training was Caltech in Pasadena, across the country from upstate New York, because I found out that Professor Richard Roberts was developing a combinatorial platform that could be used to analyze RNA-protein interactions — just the things that fit my intended career trajectory.

Because of the sheer concentration of great minds on a relatively small campus, Caltech has always been known as the Mecca of the sciences. Between 1999 and 2002, I was totally immersed in reading and learning about the fascinating word of combinatorial chemistry, from the origins of the biological world to the design of nano devices and everything in between, and yes, with RNA at the center of many things, just as I would like it.

Professor Roberts' laboratory was in the Braun Laboratories of Biological Sciences. My first experimental model was the N protein from Bacteriophage Lambda that recognizes a regulatory RNA element called BoxB RNA. Lambda is the first organism that molecular biologists and geneticists have obtained a complete molecular understanding of — well, almost complete except for the mysterious N protein. Rich's invention which attracted me to his laboratory was a platform called the mRNA Display. You could use this platform to build a high complexity library of proteins such that each member is covalently linked to the mRNA which codes for them, and this was done by hijacking the protein translational machinery. After subjecting the system to some pressure of functional challenges, the physical linkage then allows you to recover the genetic information and amplify the survivors. This gives you a system that can undergo accelerated molecular evolution — diversity, functional selection, and amplification, all done in Eppendorf tubes.

To cut a long story short, my colleagues and I played with the library construction design and selection strategies against various related RNA targets and came up with a set of N protein variants. The next obvious task was to dissect them with chemical and biological tools. As a biophysical chemist, physical chemical approaches including NMR, circular dichroism, isothermal calorimetry, and gel mobility shift were some of the obvious choices. Short of arriving at a full structural model at atomic resolution, all the biophysical evidence pointed to somewhat different modes of binding between the variant N proteins and the boxB targets, compared to the wild type structure. These observations raised a lot of questions along the lines of molecular recognition, but how do we resolve them? Rich brought me and our data to meet Dr. Zewail.

Prior to that meeting, I had rarely visited the Arthur Amos Noyes Laboratory of Chemical Physics, even it was only a few steps from Braun. The air inside these two buildings felt different for some reason — maybe the minds of biochemists and physical chemists produce different thought patterns and thus radiate different types of energies into the hallways. I knew very little about Dr. Zewail before that meeting, but at the very least I knew that he conducted what were regarded as the Holy Grail of experiments that any chemist would dream of; that is, to be able to watch chemical reactions as they happened.

We described the experimental systems and our observations in steady state fluorescence signal changes for a fluorophore we incorporated into the boxB RNA, which was at a position we knew the N protein would contact upon binding. That got AZ interested, and he predicted that if something was seen in our steady state fluorescence changes, some dynamics could be captured by time-resolved spectroscopy; in this case, by his femtosecond time-resolved set up. Up until that point, the words "femtosecond" and "upconversion" had never been part of my scientific vocabulary.

I started to bring my samples to one of AZ's femtosecond dynamics laboratories in the basement of Noyes. Just as AZ predicted, the N protein variants binded the boxB RNA as an ensemble of different conformational equilibria among substates, and these substates could be characterized by different decay dynamics of the fluorophore. The conformational complexity of these systems would have made crystallography or NMR less revealing — an average image or a single frame of the ensemble would not be a fair description of the system, but the complexity was perfect for the femtosecond dynamics approach: the N protein/boB complexes revealed themselves as conformational ensembles in front of the fast camera — so fast, all the conformational substates were frozen and waiting to be interrogated.

The initial success triggered much shuttling between Braun and Noyes, yet the difference between the air in Braun and the air in Noyes somehow never changed. While walking those short distances between the two buildings, I would have to consciously switch hats and vocabularies in anticipation of talking to different types of colleagues. Although I had always been interested in both biochemistry and physical chemistry as I was growing up, it was these short walks on the Caltech campus that made me feel like I was crossing a bridge — one that connected the worlds of biology and physical chemistry. The frequent trips between the two buildings lasted a while before I switched to AZ's group under his formal mentorship while still maintaining close ties with Rich's laboratory. That was a unique privilege which also served as a model for my independent career later. Specifically, this was the type of experience that I would want my future students to benefit from as well.

For a young aspiring scientist like me, still searching a dream to pursue at that point, any compliment from a world-renowned scientist like AZ would mean a great deal. I did not deserve too many of them during my years there, but one of my favorites from AZ that made me feel so good was at a large group meeting when I asked some questions during a presentation by another colleague, and AZ said of my questions: "Mm… that was not too bad for a biologist."

AZ had a very different style of drafting manuscripts compared to other scientists. You typically wouldn't know when he would call for a team meeting to discuss a manuscript, so you always had to be ready with your story about the data. He would neither ask you to email your draft to him nor send his comments back to you via email, which would usually have led to the typical email merry-go-round. Instead, he wanted to talk to you in person. Once he sat down with you and got your story, he would turn your manuscript into a gem with simple touches. One could only dream of having the sort of insightful intuition and perspective he had which enabled him to tell these stories in the best way possible.

After I moved to the University of Texas at Dallas to start my independent career in 2005, the only chance I would get to see AZ in person was pretty much just the annual Welch Conference on Chemical Research in October in Houston. In my opinion, the Welch Conference is the best conference because every speaker is either already a Nobel laureate or someone of that calibre. But that was secondary to my wanting to see AZ once a year, even if only briefly, so that I could absorb his infectious love for sciences.

My long term career goal is to decipher the general principles of molecular recognition. After all, all biological processes and regulation mechanisms come down to these principles. I wanted to use the ultrafast dynamics approach as my main key to unlock these mysteries, which is of course still RNA-centric. Somehow, I managed to convince at least one university to fund a femtosecond spectroscopic laboratory so that I could continue establishing RNA ultrafast dynamics as a new field.

There were times when I thought: perhaps I was building a bridge between biology and physical chemistry, as I recall those moments shuttling between Braun and Noyes. But such a thought would have been badly misguided. AZ was that bridge; he was already there. I am only happy and lucky to be a small brick on that beautiful bridge.

Photo 37.1. Members of the Laboratory of Molecular Sciences; photo taken in December 2003.

Dr. Tianbing Xia was born in 1968 in Beijing, China. He went to Peking University between 1987 and 1991 for his undergraduate education and obtained his B.S. after completing a thesis on polymer chemistry. He stayed on for another three years and received a M.S. in physical chemistry with an emphasis on the crystallography of organometallic complexes which mimic the core structure of metalloproteins. He moved to the U.S. in 1994 to pursue his graduate studies at the University of Rochester and received his Ph.D. in biophysical chemistry in 1999. He then moved to Caltech for his postdoctoral training before securing a faculty position with the Department of Molecular and Cell Biology at the University of Texas at Dallas in 2005. At the University of Texas at Dallas, he pursued a line of research in ultrafast dynamics on the probing of the RNA conformational landscape. During his more than 25 years of scientific research experience, Dr. Xia has authored, with some incredible colleagues, about 50 peer reviewed scientific publications in topics ranging from physics to chemistry to biology. In 2012, he was presented with the opportunity to work at Abbott Laboratories, a global healthcare company. He is currently a principal systems engineer with the Diagnostics Division of Abbott Laboratories, where he has been supporting clinical chemistry analyzers and developing creative solutions for customers.

38 From México to Pasadena and Back: The Pleasure and Honour of Knowing Ahmed Zewail

Jorge Peón*

I had the enormous privilege of joining the Zewail group as a post-doctoral scholar in 2001. Sometimes in life one gets lucky and many helpful circumstances arise. In my case, I was fortunate that a research paper which was submitted to the *Journal of the American Chemical Society* from my Ph.D. research group was reviewed by Ahmed Zewail himself. The review of our manuscript was full of his typical insights about the femtosecond dynamics of molecules. A few months later, when I sent one of the most important letters of my life, a letter asking to join the Zewail group, the manuscript that Ahmed had reviewed months before left a strong enough impression on him that it tipped the balance in favor of him accepting me into his highly sought-after group.

Once I arrived at Caltech in the spring of 2001, I was just amazed by the dynamism of the Zewail group. There were about twenty members (mostly postdoctoral fellows) and six laboratories each with its own femtosecond laser system. But I was most impressed by Ahmed. From the first meetings I had with him, it was clear that he

* Postdoctoral Research Fellow.
Email: jpeon@servidor.unam.mx

was a class act. It was also clear that he was able to focus on the most relevant aspects of each group member's research objective. During my time at Caltech, we worked on the fundamental photophysical properties of molecules in solution. One project was related to the ultrafast relaxation dynamics of DNA bases and the second project revolved around a basic biochemical problem: the solvation of proteins. Thanks to the perspective and direction provided by Ahmed, both projects produced important research papers which are still highly relevant fifteen years on.

Several years later while I was already a professor at the National University of México (UNAM), I received a great email from Ahmed. In the message, he said that the Caltech laboratories and equipment were being renovated and asked if I would be interested in receiving an amplified laser system for my laboratory as a donation from the Zewail group and Caltech. This equipment could double my research capacities, so of course I said yes immediately. The amplified system was the same one I had worked with in the Noyes laboratories at Caltech and, after some realignment, it was in excellent condition. That laser has since produced many more research articles at UNAM. I am very happy to say that even today, more than ten years after the donation, several parts of the system such as the femtosecond oscillator are still in use at my laboratory. I am extremely grateful to Professor Zewail for this important donation which not only increased my research productivity in Mexico but also enabled many Mexican graduate students a chance to use state-of-the-art research equipment. I am also profoundly grateful to Spencer Baskin and the personnel at Caltech for helping in the complicated logistics of the donation and transportation of the system to Mexico City.

Ahmed Zewail's passion for global issues and the progress of developing countries is well known. We in Mexico had two great opportunities to serve as hosts of Ahmed and hear important speeches on these issues. In 2004, Professor Zewail received the Honoris Causa Doctorate Degree from the Autonomous University of the State of Mexico. A year later, he received the Albert Einstein prize which, luckily for us, was awarded in Mexico in that particular year. At both events, Ahmed gave incredibly insightful speeches about the importance of investment in basic science in developing countries. I am sure that the decision makers at both events shaped their opinions significantly from the clarity of his words and writings.

I will forever be grateful to Professor Ahmed Zewail first and foremost for all his scientific contributions — sometimes I read them in articles and books; other times I had the privilege of hearing about them directly from him. Second, for the opportunities he gave me by inviting me to join his research group and for his support of my laboratory, which in turn has greatly benefited many physical chemistry students at UNAM. Lastly, for his important work with concrete proposals for the paths that developing countries should take in order to progress. The photo included in this chapter shows Ahmed Zewail together with some of the students from the Laser Spectroscopy Laboratory of UNAM's Institute of Chemistry. The photo was taken during his last trip to Mexico at Palacio de Bellas Artes in Mexico City.

Photo 38.1. Professor Ahmed H. Zewail and members of the Peon group at UNAM in one of his visits to Mexico City.

Jorge Peón was born in Mexico City in 1969. He studied chemistry at the National University of México (UNAM) and obtained his Ph.D. in physical chemistry from Ohio State University under the direction of Professor Bern Kohler. Dr. Peón joined the Zewail group as a postdoctoral scholar in 2001 where he studied the femtosecond dynamics of protein solvation and the ultrafast electronic relaxation of DNA bases. He became a faculty member at the Institute of Chemistry at UNAM in 2003. His research interests have involved the study of the fundamental photophysics of nitrated molecules, ultrafast electron and proton transfer reactions, and more recently, the design and study of multi-chromophoric systems. Dr. Peón was named Director of the Institute of Chemistry in 2014.

39 An Ultrafast Life in Femtoland

Yuhong Wang*

I did my postdoctoral training with Dr. Zewail between 2002 and 2004. I had, at that time in 2002, just graduated from Johns Hopkins University, a few publications, and a highly simplified and optimized view of research where I thought I could solve any problem with hard work. The training I received in Caltech taught me the real meaning of doing research, which is most importantly to find the right question to work hard on. I had the opportunity to work with colleagues from all over the world, and under the great leadership of Dr. Zewail, their work defined the frontiers of human knowledge. I cannot express well enough how proud I am to have once been a member of "femtoland" and how grateful I am for the experience in Caltech. It shaped my understanding of research and set the foundations for my success later in academia.

I owe it to Dr. Zewail's generosity that I could become a femtolander. My husband was accepted into his group first and after speaking with him, Dr. Zewail learned that I was also looking for a job. So he invited the both of us into his prestigious group at the top institute that is Caltech. You can never imagine how excited I was! I still remember that my hands were shaking from excitement when he called me personally to extend the offer. He jokingly asked if I was

* Postdoctoral Research Fellow.
Email: ywang60@uh.edu

good enough, and I excitedly replied that I was even better than my husband. This was the beginning of my journey in femtoland.

Dr. Zewail was very busy, but whenever he was around, he'd call for group meetings every week. There were two things that he always wanted to instill in us. One is that we shouldn't do things just because we can, but instead to do them because they are "important." The other is that he was always very nervous about the integrity of data and the potential for errors, be it human error or instrumental error. I still remember how he emphasized the term "nervous" by saying it louder and slower. Now, as a scientist, I would ask myself every day: "am I doing the things I do because they are important and not just because I can use the fancy instruments?" or "what else should I do to make the data more solid?" I can't say that I'm always satisfied, but I am at the very least trying my best to address these very important issues because they are now a part of my basic research habits.

I became a complete fan of Dr. Zewail while in the process of publishing my first paper with him. Of course, I had always looked up to him prior to that, but it was not until I worked with him on the manuscript that I gained a real appreciation for his extraordinary ability to cut through complicated data with simplicity and elegance. According to an old Chinese idiom, he "painted the dragon's eyes." The idiom comes from a story that was told 1,500 years ago. In the story, there was an artist who painted dragons but he always left out the eyes. Bystanders urged him to add eyes to his paintings, and one day he relented. However, once the eyes were added, the dragons flew away into the sky amid a sudden thunderstorm. That was how I felt after those discussions with him about my manuscript. I was working on the myoglobin-oxygen recognition, and we started from the O_2-bound myoglobin and triggered the rebinding via laser-induced dissociation. Several mutations around the binding pockets were introduced one at a time. We observed uniform rebinding in the picosecond regime but the kinetics diverged in the nanosecond regime. I worked several months trying to fit together the kinetics with different schemes and finally came up with a model that I was satisfied with. When I came up to show Dr. Zewail the findings, he complimented me first and said that these were "delicate" experiments, but then he said in a firm tone, "I do not believe the numbers." The data had been fitted with too many parameters. He then came up with a streamlined explanation which was eventually illustrated in Figure 5 of our paper (Wang, Baskin, Xia, & Zewail, 2004; published in *Proceedings of the National Academy of Sciences of the United States of America*, see Photo 39.1). In this figure, Dr. Zewail elegantly interpreted the data with a protein bifurcation model without getting entangled with trivial, specific numbers. My eyes were opened from the realization that I was distracted and drowned with these trivial numbers. Once he pointed it out, it became so simple and obvious. I was so inspired from this experience and I suddenly understood why Dr. Zewail was a Nobel laureate. Looking far and deep was second nature to him. Till this day, I make it a point to remind myself not to get overly engrossed in trivial details which may obscure the bigger picture.

The bifurcation model also reflected Dr. Zewail's vision about biomolecular dynamics, which he shared with us multiple times during group meetings. His opinion is that although biomolecules move at relatively slower time scales (milliseconds to seconds), the fate-determining dynamics must occur with a sharp transition at much shorter time scales, even in the femtosecond to picosecond range. These two types of dynamics exist in bifurcated populations that are coupled to each other. This view also helped me to understand "coherence" better, another key concept that I learned from Dr. Zewail. On the one hand, from the point of view of time, the ensemble average effect occurs whereby molecules are "in phase" for a very short period of time before coherence is lost to complicated processes. On the other hand, coherent populations may be retained if the molecules can be tracked individually using single molecule approaches. Following this idea, I started my second postdoctoral stint learning about single molecule approaches. I then started my own independent research group at the University of Houston in 2008. My very first independent paper (Altuntop, Ly, & Wang, 2010) was about observing and sorting the heterogeneous subpopulations of ribosome, the protein-making factory, using the single molecule FRET approach. The model that I proposed was inspired by the *Proceedings* paper mentioned above. This paper was the critical publication that established my tenure five years later.

The two years I had at femtoland were ultrafast because it was a fateful and life-changing period for me. Before I went to Caltech, my life goals were not clear. I was just basically progressing through the grades in the schools I attended, one after another. At Caltech, Dr. Zewail presented me with both the resources and the challenges. There were no more clear steps to follow. After a short period of confusion, I got on track and enjoyed my time at Room 036 in Noyes building. After two years at Caltech, my life goals became clear and I knew that I wanted to be a professor. And now, I am a tenured professor at the University of Houston under the Department of Biology. Today, we are here to celebrate Dr. Zewail's life. I am grateful to him every day for giving me my academic life and purpose.

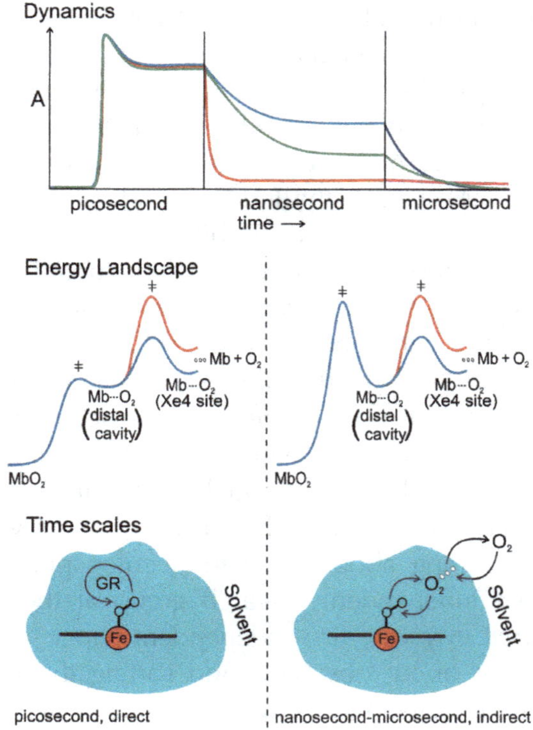

Photo 39.1. A streamlined explanation of the kinetics of myoglobin-oxygen binding using the bifurcation model as presented in Wang, Baskin, Xia, and Zewail (2004).

Photo 39.2. A group photograph of me (second from the left in the first row) and the femtoland group.

Yuhong Wang is an associate professor at the University of Houston, Department of Biochemistry. Her research is focused on the study of the mechanism of biomolecules using biophysical and biochemistry approaches, in particular the fidelity mechanisms of ribosomal protein synthesis, as well as the development of new biophysical methods.

Yuhong Wang obtained a B.S. in chemistry from Peking University and a Ph.D. from Johns Hopkins University in 2002 by studying the time-resolved Infrared spectroscopy of carbenes. She then studied myoglobin-oxygen recognition using ultrafast flash photolysis under Dr. Zewail's advice. Before becoming a faculty member at the University of Houston in 2008, Yuhong continued her second post-doctoral research learning single molecule techniques at the University of Pennsylvania. In her spare time, Yuhong enjoys baking, watching movies, and wine tasting.

40 A Capturing Moment with AZ

Amisha Kizhakkedathu*

In January 2003, I relocated with my newlywed husband from India to the United States to join AZ's laboratory at Caltech as a postdoctoral researcher. We were received at the airport by Ramesh Srinivasan, a Ph.D. scholar from AZ's laboratory. A temporary accommodation at the nearby Vagabond Inn was already arranged by AZ's office for us. We were so wonderfully received in the beautiful city of Pasadena that it felt like a perfect honeymoon. I was impressed at AZ's thoughtfulness.

In the coming days, I would work very hard in one of the several "femto" laboratories, setting up the femtosecond laser and shooting this laser on protein samples to probe water dynamics during enzyme activity and protein folding and unfolding. Within a year, in January 2004, we produced one paper titled "Enzyme functionality and solvation of subtilisin carlsberg: from hours to femtoseconds" in the journal *Chemical Physics Letters*. During those days, AZ used to travel around the world a lot and we had to fully utilize the limited time he was available in the office to have our scientific discussions and wrap up our manuscripts. My fellow postdoctoral colleague

* Postdoctoral Research Fellow.
Email: amishakamal@gmail.com

Liang Zhao and I would be in Laboratory 7 (there were a total of seven laboratories in AZ's group if I am not wrong) even on the weekends to set up experiments. We took turns to use the laser and lead our own experiments.

I thoroughly enjoyed my time there. I studied in-depth the decay behavior of intrinsic and extrinsic fluorescence of proteins in femto- to picosecond time scale during enzyme activity and protein folding/unfolding and correlated it with the dynamics of water molecules in the protein's hydration shell. Around the time I completed the second project titled "Ultrafast hydration dynamics in protein folding/unfolding: human serum albumin" and was ready to discuss it with AZ, his schedule turned out to be very tight. He emailed me and said that he could discuss the data and wrap up the preliminary manuscript that I drafted the following Saturday. After that, he would be gone again for a couple of months or so.

I immediately cashed in on that suggestion. We met on a Saturday in May 2004, discussed the data in the first half of the day, and wrapped up the manuscript for submission to the journal *Proceedings of the National Academy of Science* in the second half. Research administrator De Ann Lewis was also in the office to provide administrative (as well as moral) support. I wanted to capture this special moment to be cherished forever and I got a passerby to help us take a very special photograph (Photo 40.1). Whenever I gaze at this picture, I can't believe that AZ is no longer here with us. He greatly embodied charisma, energy, and passion for science! May his soul rest in peace.

Photo 40.1. AZ and I working on a manuscript on a Saturday in May 2004. From left to right: Amisha Kizhakkedathu, Ahmed Zewail, De Ann Lewis (research administrator).

Dr. Amisha Kizhakkedathu is currently an associate research fellow/ senior group leader in the bioanalytical development group within the biotherapeutics division at Pfizer. In her previous roles, she was leading the analytical development group within biotherapeutics division at Astellas, and method development and characterization group within biologics division at MedImmune/AstraZeneca. Prior to these roles, she was a Sr. Scientist within Biotherapeutics at Pfizer after completing a 4-year Research Scientist role in the laboratory of Mass Spectrometry pioneer Dr. Mark Chance at Albert Einstein College of Medicine, Bronx, and Case Western Reserve University School of Medicine, Cleveland. She obtained her Ph.D. in biophysics from India's premier research institution, the Tata Institute of Fundamental Research, Mumbai. She then did her postdoctoral research under Nobel laureate Dr. Ahmed Zewail at California Institute of Technology, Pasadena, as well as in the laboratory of Dr. Chance.

41 A Decade with Dr. Ahmed Zewail and Beyond

Ding-Shyue (Jerry) Yang*

California Institute of Technology (Caltech) may very well be called my second home. A whole decade passed as I completed my Ph.D. and postdoctoral work there, maturing academically with the guidance, help, and patience of Dr. Ahmed Zewail. During the same period, I got married, received the births of two lovely daughters, and built a family. Each time I reflect upon my time at Caltech, I am thankful for the lasting impact that Dr. Zewail had left on me.

I knew that joining the Zewail research group would be the best choice I could have ever made as I left my home country, Taiwan, for Caltech in 2002. However, I was intensely worried about my limited command of spoken English at that time and it took me quite a few months before I mustered the courage to ask for a meeting with Dr. Zewail. It had been only three years since he received the Nobel prize and I figured that better-prepared students would have been accepted into his group over me. However, he was kind enough to spend some time talking to me and in a few hours, he informed me that I was accepted with a warm welcome at a Friday evening group meeting. Only some time later did I find out that the opportunities for students were in fact quite limited, and I was very fortunate to be among the last few graduate students he took.

* Graduate Research Assistant/Postdoc.
Email: yang@uh.edu

It was an invaluable opportunity and experience to be able to witness for an extended period of time how Dr. Zewail assembled talents and guided each research direction with his vision, enthusiasm, and determination. At the same time, he did not compromise on the little details and instilled in me the discipline of scientific rigor. Coming from a theoretical background without much prior experimental experience, I spent my first year of research trying to catch up and absorb as much as possible alongside Hern Paik, then a graduate student. I began to learn about the level of preparation that was required before I was "summoned," as part of a research subgroup, from the "Femtoland" subbasement laboratory to his office on the ground floor for a discussion of the progress of our work. He would ask keenly about how we analyzed and interpreted our data while probing our reservations or thoughts about the research projects. He would be never satisfied unless we gave evidence-based responses.

After a full year in the group, a pivotal and decisive moment for my future academic directions came. One day in early 2004, Dr. Zewail called me personally and asked, to my surprise, what other research areas I would also be interested in. Apparently, he had in mind a change that I was not aware of beforehand, but I was glad that my interest in ultrafast electron crystallography (UEC) had aligned with his plan. I am indeed grateful to have been given the opportunity to contribute to the development of the UEC methodology from its early days. Between then and mid-summer of 2004, I had a period of less than four months to quickly learn everything I could about UEC from Chong-Yu Ruan, then a postdoctoral scholar in the group, before he would leave for a faculty position at Michigan State University.

However, a substantial breakthrough in my appreciation and understanding of UEC was slow to come. For about a year and a half, we conducted various different experiments hoping to make better sense of the acquired data. The rather stagnant progress (or, more suitably in my opinion, the "not-so-profound findings" so far) led to "the talk," which apparently was a common experience shared by many group alumni. In this private, long conversation with Dr. Zewail in June 2006, I felt his strong disappointment in the outcomes and as a result felt immensely pressured, but at the same time, I was also encouraged by his confidence in me that I would have the ability to eventually make things right. Like what many others may attest, he knew well the art of managing his group members and pushing our limits. I was glad and relieved when I redeemed myself a few weeks later with some good results and satifactory understanding, and did not make him wait too long.

Dr. Zewail was excited about my presentation of the UEC results on gallium arsenide. In the summer of 2006 in the backyard of his house, he was pleased that I was able to put together a good physical picture for the experimental observations and answered with confidence all of the questions he posed. As a graduate student, receiving praise from a Nobel laureate made me feel like I was on top of the world, especially when he acknowledged that he learned from these discoveries together with me. However, in the following weeks, I was soon brought back to reality; this time, the trouble was with my writing.

It was already the end of my fourth year in the Ph.D. program but I still did not have a firm grasp of how to effectively communicate my ideas and findings in scientific papers. I can still vividly remember his frustration as he tried to work on the manuscript I gave him, and its extended length as a full article certainly did not help. Several times, he said "maybe we should stop" or "should we continue?" with a struggling facial expression. He could have easily pulled the plug and called it off, and there were indeed a few close calls. Thankfully, he went along with my attempts to calm him down and bore with me for a few more months. The whole process was atypically long and unlike the usual effective, concentrated pace he would go at once he had set his mind on a particular manuscript. But because of his patience, I learned a lot and managed to develop a better understanding of the word choice and presentation structure required in manuscript writing. This experience was critical to my academic training and career development.

There were also other moments when he teased me, but they were often for my own good. In November 2007 after a group meeting, Dr. Zewail lightheartedly joked that I could not stop using "basically" in almost every sentence I said, to the point that it had become distracting. I was totally unaware that I was repeating this filler word over and over again! And this comment of his has stuck with me ever since. On 1 May 2008, I presented an Everhart Lecture to a campus-wide audience with him in attendance, something which I would never have imagined to happen given my intense anxiety of presentations in English during my earlier years. After the seminar, he gave me an approving smile congratulating me on my much improved oral presentation skills. Of course, he did not forget to poke fun at me and said, "I am surprised you did not say 'basically' even once in your talk!"

My Ph.D. years concluded in 2009 after almost seven years at Caltech. With Dr. Zewail's full support, I stayed on to work on the design and instrumentation of a new time-resolved electron imaging technique, the so-called scanning ultrafast electron microscopy (SUEM), to visualize the photo-induced dynamics of materials in real-space scan images. I am proud to be able to say that I was the one who put together this (in-field) proposal as a new project, considered the details of its feasibility, and then carried it out together with Omar Mohammed, then a postdoctoral scholar. Thankfully, we achieved our breakthroughs on the project within two years and I could move on to the next stage of my career with Dr. Zewail's blessing.

During my postdoctoral period, it became harder to interact with him as frequently as I did in previous years. Besides the research frontiers of four-dimensional electron imaging, he had additional major responsibilities including being part of President Obama's Council of Advisors on Science and Technology from 2009, serving as the United States' First Science Envoy to the Middle East from the same year, his involvement in helping his homeland during the Egyptian revolution between January and February 2011, and the creation of the Zewail City of Science and Technology also from that year.

Knowing that I am a person of Christian faith, Dr. Zewail once said to me, "I am a person of faith, too." Although we came from different nations and had different cultural backgrounds and heritages, I have always admired his strong dedication not only to "pushing the frontiers of science" as he often said, but also to fighting for the resolution of various conflicts and pressing issues in America and around the world. Certainly, he was a person of faith who pushed for a better world with peace and welfare, especially through the advancement of science and technology as well as the encouragement of dialogues among people with contrasting views.

I have also hoped for a better world, practicing the philosophy of Pastor Jung Myung Seok for more than 22 years. Twice I had the honor of singing in front of Dr. Zewail as he knew early on that I was good at singing. During the 2005 group retreat in El Capitan Canyon, Santa Barbara, California, following his request, I sang a song titled "Life" whose lyrics were adapted from the translation of a poem by Pastor Jung. Dr. Zewail smiled and nodded after I finished the song. Twelve years have since passed, yet this same poem may continue to resolutely portray the legacy he left behind as well as his philosophy of perseverance and dedication, in particular these two excerpts:

> *When I grow into a big tree and produce leaves and flowers,*
> *I will draw butterflies and bees to me in droves.*
>
> *Go ahead and dig deep.*
> *Don't spare water or fertilizer.*
> *Make the sweat of your brow pour down*
> *Like the summer rains from your head down to your toes.*

The world will miss a great leader like him who had the capability and vision to build strong partnerships between peoples of vastly different backgrounds. I will miss my mentor for the great and everlasting impact he has had on me.

A late night talk during the group retreat on August 16, 2005. From left: Ahmed Zewail, Milo Lin, De Ann Lewis, Jerry Yang, Ramesh Srinivasan (back), Chaozhi Wan, Hern Paik, Sang Tae Park, and Jon Feenstra (foreground).

Dr. Ahmed Zewail and me in our regalia in his office after the commencement ceremony on June 12, 2009.

Ding-Shyue (Jerry) Yang received his B.S. and M.S. in chemistry from National Taiwan University in 1997 and 1999, respectively. Under the supervision of Dr. Ahmed Zewail at Caltech, Jerry was involved in the development of ultrafast electron diffraction for condensed matter and completed his Ph.D. in May 2009. He continued on as a postdoctoral scholar and a research scientist in the Zewail group and successfully designed and built the first scanning ultrafast electron microscope. In 2012, Jerry began his independent research at the Department of Chemistry at the University of Houston. He was a recipient of the junior Welch Professorship in 2013 and the National Science Foundation CAREER award in 2017. Jerry's research interests include ultrafast dynamics and transformation pathways of transition metal materials, interfacial energy- and charge-transfer dynamics in heterogeneous solid–molecule systems, and the development of electron imaging methods.

42 Voyage through Femtoland

I-Ren Lee*

It was a wonderful experience working in Femtoland under the supervision of Dr. Zewail. I still remember when I first came to Caltech in June 2003 with my very broken English. He was always very patient, listening to me and making sure we were on the same page when he gave me advice. Later, he mentioned multiple times that passion and patience are the key components of doing good science. I realized that he was practicing this throughout his sparkling research career.

Hours after joining his group, I was assigned to the Noyes 048 laboratory. I guess the main reason for this assignment was that my former master thesis advisor in Taiwan, Professor Po-Yuan (James) Cheng, had worked in the same laboratory and it was good for me to continue the heritage. I was working with Dr. Daniel (Hern) Paik and Dr. Ding-Shyue (Jerry) Yang, who were both also graduate students at that time. Since we were the only subgroup still doing gas-phase femtosecond spectroscopy, most of our projects were extremely difficult to conduct experimentally and it took months just to get the experiments started. However, Dr. Zewail never gave us undue pressure. Instead, he was very patient and always encouraged us by reminding us about the significance of the projects and telling us that

* Graduate Research Assistant/Postdoctoral Fellow.
Email: irenlee@ntnu.edu.tw, or irenlee@gmail.com

we would be famous if we could make it happen. For me, it felt like we were sailing on the ocean at night and you could see nothing except the lighthouse that was Dr. Zewail, as he guided us through the hard times. Many times, after several months of effort, we would finally get some exciting results. I could always feel his excitement and he never failed to tell us that he was proud of us.

Although Dr. Zewail usually looked very serious, he had a great sense of humor and he also liked it when we joked around from time to time. Since Femtoland is located in the subbasement and we usually finished our work after dusk, we didn't usually get sunburns. But when we did get sunburnt from ditching the laboratory and doing outdoor activities, we would always joke that we got burned by the ultraviolet radiation when working on the frequency tripler. He laughed heartily and we all very much enjoyed that moment. I also often cracked jokes about my weight. He was a good observer, so it came as no surprise that he noticed my body weight "oscillated" (changed up and down) quite frequently, just like a wave packet. I tried to come up with an amusing reason for this, so I told him that when I gained weight, it was because I didn't have time to exercise and when I lost weight, it was because I didn't have time to enjoy my meals, and this was all due to my hard work in the laboratory. He laughed together with Ms. De Ann Lewis (his project administrator) and we all had a good time. Working in Femtoland was never easy and we all were under tremendous pressure in the pursuit of success for our careers. His good sense of humor helped ease that tension and made the atmosphere very pleasant.

I especially enjoyed the time we wrote a paper on the lawn in the backyard of his house under the warm California sunshine. He always made some good tea while he spent his weekend with us working on our papers. In most cases, he would rewrite the introduction section nearly completely, taking the paper to a different level. We could only admire his magic. At the end of the day, the paper would be nicely done and ready to be submitted on Monday. It was always so efficient working with him and he always encouraged me by saying that if I kept working hard, one day I could be as good. I will keep trying to get better even though my ceiling may not be as high.

Dr. Zewail was no doubt very passionate about science. However, he never lost his awareness of public issues such as politics, religion, and humanity. During a group retreat, we sat around a campfire smoking a cigar and exchanging our opinions. It was fun to listen to the different opinions of people from varied cultures, backgrounds, and countries. He sometimes reminded me not to get myself too involved into politics, but when the Egyptian revolution happened in 2011, Dr. Zewail stood up, put himself in a risky position, and fought for the good of his people. After the revolution, he spent a lot of time building Zewail City. His selfless devotion to his motherland inspired me to make contributions to my own country.

I benefited a lot from the solid training I received in his group. After leaving, I always missed the good old times in Femtoland. The image of an energetic superhuman remains my fondest memory of Dr. Zewail. Unfortunately, he couldn't overcome his illness and left us forever, too soon. Yet, his spirit always lives on in our minds, continues to oversee us, and guides us towards doing good for humanity. Dr. Zewail, you will be forever missed.

Photo 42.1. I was lucky to have Dr. Zewail being the commencement speaker of 2011, the year I graduated. This photo was taken after the ceremony, behind us stands Professor Aron Kuppermann.

Dr. I-Ren Lee joined Dr. Zewail's group in summer 2003 from Taiwan. He first worked in the Noyes 048 lab doing femtosecond spectroscopy studies on anionic species. In 2009 he moved to the ultrafast electron diffraction (UED) lab working on UED of biologically relevant and thermally labile molecules. In 2010, he passed his thesis defense and stayed in the same lab as a postdoctoral fellow for another year. Then he moved to the University of Illinois at Urbana-Champaign for his postdoctoral research and subsequently obtained his current position, assistant professor in Chemistry at National Taiwan Normal University in 2014.

43 Seeing Beyond My Vision

Roberto A. Garza-López*

I first met Ahmed Zewail when he visited our department at Pomona College. He was our Robbins lecturer in 2002, which was 40th in the series established by Mr. Robbins to bring distinguished chemists to Pomona College to discuss their current research. This lectureship allows us, the faculty in the chemistry department, to annually invite a scientist who has won the Nobel Prize or is at the top of his or her area of research. We wanted Ahmed to be one of our lecturers and we worked very hard to lure him to Pomona. It was not an easy task especially when you are talking about a famous figure with a very busy agenda, but *we did it!*

Ahmed delivered a series of talks that he titled "Seeing Beyond Our Vision" and his visit was an amazing success. It started with a dinner which saw not only the participation of Pomona students, faculty, and staff, but also nearby high school students and their teachers. This first dinner was an opportunity for the audience, specially students, to ask Ahmed about the Nobel Prize and his scientific discoveries, as well as other fun personal questions such as the courses he liked or disliked when he was an undergraduate or high school student, his thoughts about how to achieve success in the sciences, and how the Nobel Prize changed his life. I remember that

*Visiting Associate.
Email: RAGL4747@pomona.edu.

Ahmed was thrilled to be able to dispense jewels of wisdom as he answered the many questions he received, and I now wonder just how many of those participants' lives had changed for the better because of those interactions with him.

I know for a fact that Ahmed improved my life tremendously. After dinner, it was time for his first lecture. The first talk was geared toward the public, the auditorium was packed full, and after he finished, the audience bombarded him with questions which he was only too eager to entertain with that characteristic grace he always exuded. Ahmed's charisma attracted audiences composed not only of scientists but also of non-scientists eager to learn how science affects their lives, and both groups were well served for he was not only an outstanding scientist but also a great communicator. He was a titan with a unique mind and a gentleman who shone a light so bright, all the awesome possibilities of the universe would become illuminated for everyone to see. He had that ability and charisma to sell science to anyone and he always made an effort to deliver his scientific lectures just like how a brilliant actor would deliver a stellar performance and captivate the audience. The other lectures he gave were more "scientific/technical" but were nevertheless still a treat to the many science faculty from the Claremont Colleges who eagerly attended.

More dinners followed where Ahmed was able to delve into his academic adventures, science, politics, his family, his life at Caltech, and his trajectory from his initial studies in Egypt (his homeland) to receiving the Nobel Prize. He certainly enjoyed his visit; we all were thrilled with every lecture he delivered and every conversation he had with the many students, faculty, and people who came to hear him speak. At the luncheon with the members of the chemistry department at Pomona College, I remember him talking about issues he cared deeply about — funding for basic scientific research, the importance of science, how to connect with non-scientists, and how to help underdeveloped countries through science. He was a humanitarian and felt that he had a responsibility to help others through his reputation and influence. This motivated me tremendously. Here was a Nobel laureate, a scientist known throughout the world who not only loved science, but who also worked tirelessly to promote his idea that developed countries like America had a responsibility to help those who are underdeveloped by sharing talent, expertise, and technology. I could say that I was captivated not only by his brilliance but also his care for the disadvantaged. He was certainly a famous scientist with a big heart.

In one of my conversations with AZ, we shared our experiences as scientists from countries like Mexico and Egypt who came to America to fulfill the American dream. I remember that he was asking me many questions about my research when the idea of me doing a sabbatical at his laboratory came up. He was very enthusiastic about the idea and the realization that this was a possibility was like a dream come true for me. We then talked at length about possible research problems to tackle and eventually settled for two main directions: energy transfer and the molecular dynamics of proteins.

The research experience I had at Caltech was fantastic and I was amazed at how he kept abreast of the different research lines undertaken by his many graduate students, postdoctoral scientists, and visiting professors in his laboratory.

Even though chemical research was the initial objective of my sabbatical, another idea began to emerge after reading his autobiography, *Voyage through Time: Walks of Life to the Nobel Prize*, which was published in English. Since I help and sponsor economically disadvantaged young students in Latin America, I also recognized that the powerful message expressed in his autobiography could be usefully transmitted to young minds in the underdeveloped countries of Latin America. When I proposed the idea of translating his autobiography to Spanish, I could see the characteristic enthusiastic glow on his face. He was thrilled with the idea and gave me the green light to proceed. I knew that the process was not going to be easy, but Ahmed and I knew it would be immensely beneficial to all the students in underdeveloped countries who could change their lives by embracing the study of sciences, and thus it would be worth the effort. Embodying the title he chose for his Robbins lectures at Pomona, Ahmed was indeed someone who always "saw beyond his vision." After all, what better example could there be than his own life story to motivate those young minds? His was the story of a boy from Egypt who had committed himself to drinking from the fountain of scientific knowledge, went to America, got his Ph.D., and secured a faculty position at one of the most prestigious universities in the world. In a few short years after he made the discovery and became the father of femtochemistry, his unique contributions to science were recognized when he was awarded with the Nobel Prize in chemistry. For discovering the "fastest camera in the world" and starting a new branch of chemistry, Ahmed became the first Arab to obtain the Nobel Prize in chemistry unshared.

I wanted the autobiography to be published by the best editorial house in my native country, so I contacted the representatives of El Fondo de Cultura Económica which was by far the best in Mexico and Latin America. I submitted my proposal and a selection committee comprising 22 faculty members (all with Ph.D.s in Spanish language, literature, and culture to advise the editorial) enthusiastically accepted it. When I communicated this to Ahmed, he was delighted because he knew that this was an opportunity to transmit his message to young scientists who didn't have the resources found in America, but who could still dream of a better life and make those dreams a reality.

The first obstacle which took some time to resolve was for the Egyptian editorial to sell all the exclusive rights to the Mexican editorial to publish Ahmed's autobiography in Spanish. I couldn't wait that long, so I just went ahead and immersed myself into translating each chapter and contacting several of my friends from Spain and various Latin American countries to make sure that some Spanish terms I used meant the same thing across different Spanish-speaking countries such as Mexico, Chile, Argentina, Colombia, Spain, and so forth. I spent my sabbatical year alternating between learning how to calculate the energetics of proteins via molecular dynamics, performing random-walk calculations on surfaces, and translating Ahmed's autobiography. Each chapter was scrutinized by the editorial advisory committee and finally the whole book was translated. The title in Spanish became *Viaje a Través del Tiempo: Senderos Hacia el Premio Nobel*. A picture of Ahmed and me (Photo 43.1) is presented on the inside cover of the book (Photo 43.2). After its publication, Ahmed received many invitations from universities in Latin America and

Photo 43.1. Professor Ahmed Zewail (Caltech) and Professor Roberto Garza-López (Pomona College) in Zewail's office.

some universities even made this book a required text for students in the sciences, mathematics, and engineering. I wonder how many minds in the "have-not" countries, as a result of reading the book, have been awakened to pursue scientific careers to better their lives and that of their loved ones, and by doing so, are on the road to making a real positive contribution to their countries of origin.

Today, I continue doing the research that I initiated in Ahmed's laboratory. I learned a lot from this great scientist and human being, and I am truly blessed to have been able to meet him in my life. When I make my way to my office or laboratory at the chemistry building of Pomona College, I often pause to see the display of pictures on one wall at the entrance lobby of the many Robbins lecturers that have visited the college. When I see Ahmed's picture, I always feel like I can hear him telling me to do my best, reminding me of my responsibilities to others, and urging me to sharpen my skills and

Photo 43.2. The cover of the Spanish version of Ahmed's autobiography, reprinted here with permission from the publisher, Fondo de Cultura Económica.

become the best scientist I can be. I can hear him telling me, "Roberto, see beyond your vision." I can clearly hear him advising me to honor the path I have chosen by contributing to science and helping others, just like he did. Needless to say, Ahmed has left a void that will never be filled. May he rest in peace.

Professor Roberto A. Garza-López began his academic career in 1992 at Pomona College in Claremont, California, and ascended through the ranks and is now a full professor of chemistry. He is a physical and computational chemist who holds a B.S. in chemical engineering from Texas A&M, a M.S. from the University of Texas, El Paso, and a Ph.D. from the University of Georgia. He chaired the department of chemistry at Pomona College for three years and was the first Cambridge University-Pomona College exchange professor. He has done sabbaticals at Stanford University in the laboratory of Professor Richard Zare, at University of California, San Diego in the laboratory of Professor Andy McCammon, and at Caltech with Professor Ahmed Zewail. He has two main areas of research. The first is the study of diffusion-controlled reactions on Euclidean and fractal surfaces as well as through zeolites. The second, in collaboration with Professor John J. Kozak from DePaul University and Professor Harry Gray from Caltech, deals with the stability of unfolding of proteins such as intelectin-1, cytochromes, and blue copper proteins. He is starting a sabbatical at Caltech in the laboratory of Professor Harry Gray where he will continue exploring the folding and unfolding of proteins related to disease via molecular dynamics simulations and other theoretical approaches.

44

My Friend Ahmed and I

Bengt Nordén*

Ahmed and I were close, or at least that is what I always felt — he called me his big brother (I was one year his senior). While I was more direct about most things, even personal matters, he was generally more difficult to read and, everybody I know agrees, was intrinsically quite private. However, we shared the same sense of humour and the same childlike curiosity when approaching scientific problems — though his was that of a genius. Ahmed was very social and always looked out for everybody, whether they were janitors or Nobel laureates. To really see the deep value in people is something I learnt from him and indeed find very important in life. His social acumen was always combined with a keen sense of humor: I remember many a good laugh shared between us. At the same time, he was reflective and a deep philosopher — often finding scholarly parallels with ancient Egyptian or Greek science.

We met more frequently after his Nobel Prize, when our families got to know each other in Stockholm. The first time we met was actually only five years before he received the Prize. Ahmed was touring in Sweden and giving seminars at various universities, and the story as he loved to tell it goes as follows. Ingmar Grenthe, my

---

*Visiting Associate.
Email: norden@chalmers.se

friend and former colleague from my alma mater, Lund University, and a previous Chair Professor of Inorganic Chemistry at KTH Royal Institute of Technology in Stockholm, one day called me and suggested that I host Ahmed Zewail for a seminar at Chalmers. I said, "Zewail who?" and Ingmar became quite upset and asked how I could so be ignorant of the most famous laser spectroscopist in the world. My ignorance turned out not to be true though, although I was confused at that moment: as Ahmed and I found out when we met in my office some weeks later, and I pulled the Swedish National Encyclopedia from the shelf, there was a several pages long article authored by myself some five years earlier titled "Physical Chemistry," featuring molecular reaction dynamics studied by fast laser spectroscopy. It contained an illustration from one of Ahmed's papers! Further, according to Ahmed and Ingmar, I should have said that he (Zewail) would be warmly welcome provided costs for his travel to and stay in Gothenburg were taken care of by Grenthe or the

Photo 44.1. At an Alfred Nobel Symposium on Energy in Cosmos, Molecules and Life, arranged by the Nobel Foundation at Sanga Saby outside Stockholm in June 2005. From left to right: the author, Ahmed Zewail and Ingmar Grenthe.

Academy. I translated my encyclopedia article for Ahmed, which initiated our intense discussion about where physical chemistry was heading. I would say we found each other then. I even paid for his tram ticket when we went to my home at sunset where my surprised wife, Gunnela, improvised dinner for us. Ahmed stayed long that night and we touched on many basic questions, many of which we have returned to at our encounters later at Caltech.

We got closer after his Nobel Prize partly because of our joint interest in molecular spectroscopy and, I think, partly also because of a combination of the sort of isolation that Nobel laureates often go through and me being a member (at the time Chair) of the Nobel Committee. Both of us felt a pressure to behave in a certain way — to not openly admit any of those embarrassing questions and doubts that incessantly plague scientists who want to truly understand how things really work. We both felt that the Textbook Word and reality are more often than not two different things; it is akin to a priest who might have his private doubts. Both of us believed in intuition as the best guidance to new theories and experiments, and also allowed ourselves to be secretly skeptical of many "accepted facts" in science.

I used to visit Ahmed at Caltech for a couple of days each November, and one spring I stayed for several months as an invited visiting scholar in his laboratory. I am very grateful to him and his wonderful wife Dema for their ever generous and amiable hospitality during these stays which were truly inspiring to me. The schedule of the first day of my visits always involved discussing various things that had happened in our respective research lives since we last met — these discussions were mostly dominated by Ahmed's often much more spectacular experimental progress and discoveries. We always started our day with a cappuccino together in the sun outside the campus cafeteria. There, he would ask about things related to the Nobel prize. Despite Ahmed's status as a Foreign Member of the Royal Swedish Academy of Sciences, he would be uninformed of things like who got nominated. Foreign Members can attend meeting where nominations and investigations of various candidates are presented and discussed, but they do not have the right to participate and vote at the general assembly which finally decides who the Nobel prizes should go to, and they are also not informed about any details of ongoing investigations, such as who the "hot" candidates are. Ahmed cared very much for the status of the Nobel Prize and sometimes had his concerns about true or suspected candidates. I'd always found this "updating" chat to be the most difficult to manage in our friendship as Ahmed's natural (scientific!) curiosity felt no bounds, while my own integrity and responsibility towards the Nobel institutions prevented me from divulging such information. At the same time these discussions were very useful for me as Ahmed was always an updated source of current developments, indications of upcoming breakthroughs, paradigm shifts, and so on. Our own fields were thoroughly ventilated, but any developments that were important to chemistry as a whole were also considered.

Moving from the coffee table to Ahmed's office, or rather his big conference table and whiteboard across the hallway which we filled with sketches and equations, we plunged into science. Two topics were recurrent: diffraction of singular electrons in a Young's double-slit setting and coherence of molecular vibrations. I was very much the skeptical Dr. Watson to the proud Sherlock Holmes who was Ahmed, presenting his new results like a cat presents a dead mouse at its owner's feet. He also enjoyed using me as Devil's Advocate: for example, I was generally suspicious of visible oscillations after having revealed some experimental artifacts in laser spectroscopy back in the 1980s, one being a far less impressive ringing effect due to thermal-lens fluctuations instead of the so-called Optical Kerr effect, which was erroneously claimed to be a real molecular orientation effect due to interactions between the photon field and induced molecular dipole moments. Very early on, Ahmed's elegant demonstration of recurrent wave-packet travels in the potential diagram of sodium iodide became the focus of our discussions. Was his result potentially in conflict with Heisenberg's uncertainty principle? The answer we finally agreed on was, however, "no," because the fact that Ahmed was exciting all the NaI molecules in phase would make the sample behave differently from a quantum mechanically described single molecule (microscopic system), and be more like a coherently vibrating macroscopic system.

A pedagogical device that Ahmed used in his Nobel lecture, constructed of Perspex and bent steel strings by my engineer Mr. Tore Eriksson, showed how a steel ball was moving back and forth in the excited-state potential of NaI. When the potential curve came close to the ground-state potential, the ball could jump over to the ground state when triggered by a hand-controlled mechanical switch, and then either return to the deep valley representing the bound state of the molecule or move in the opposite direction in correspondence to the dissociation into Na and I atoms. Ahmed was very fond of the device, which had an honorary position on a shelf in his office.

Ahmed's other great interest in single-electron diffraction was the embryo to his next big discovery: the use of time-resolved electron microscopy to resolve fast processes. We devoted a lot of time to discussing whether the Coulombic repulsion between electrons would create a longitudinal "anti-bunching" along the train of electrons or just spray the electrons randomly in space.

As Ahmed's guest at Caltech, I had the privilege and pleasure of meeting his friends at the "Round Table" Wednesday lunches at the Athenaeum. These are 10 of Caltech's most outstanding scientists, including Rudy Marcus and Jack Roberts. Jack passed away in October 2016 aged 98, and he was the legendary inventor of physical organic chemistry and the modern use of nuclear magnetic resonance for studying molecular structure and dynamics. To a newcomer, these lunches were indeed inspiring and also a little scary — I suddenly had to stand up and answer for all the science, politics, and economy issues that had been going on in Europe since my last visit; I generally managed to duck these killer questions although not always! At these Round Table discussions about anything and everything, I also realised how Ahmed was a born survivor. A fish in water in these situations, he would always somehow pull through, the same way he survived when he was first up for an interview at Caltech and became afraid he would misspell "Feynman" when writing it on the blackboard. He got away with it — after writing "Fe," Ahmed turned around with a charming smile and said, "We all know how to spell Feynman, do we not?" Whereupon everyone laughed.

To Ahmed, the coming generation of young scientists was very important and the youth of Egypt had a special place in his heart as will Egypt be forever grateful to Ahmed for all his influence and academic initiatives there, such as the Zewail City of Science and Technology, his lifelong dream, inaugurated with Ahmed as its Chair in 2011. He also played a seminal role in the early development and establishment of the Molecular Frontiers Foundation, chairing its first Scientific Advisory Board and speaking at many of its symposia, describing and explaining the findings at the frontiers of research in his characteristically vivid, simplistic, and pedagogical way. Molecular Frontiers owes Ahmed a lot in that he created a discussion atmosphere that today is a hallmark of the organization: the youth are prized for their questions, not for knowing the answers!

My first experience of this talent of his was at a Lindau Nobel Laureate Meeting where I was asked to moderate a round-table discussion, an occasion that gave rise to another humoros anecdote that Ahmed liked to tell. In addition to Ahmed, the Nobel laureates around the table were Harry Kroto, Paul Crutzen, Richard Ernst, and George Olah (see Photo 44.2). I received scraps of paper with questions from the floor which was composed mostly of young Ph.D. students who had received stipends to partake in the meeting. As most questions were thoughtful and scientifically advanced (read: boring), I sneakily added in a few questions that fired up the discussion considerably. One such question asked, "With reference to Alfred Nobel, how does dynamite really work?" Relatedly, another question asked, "What is the role of the stabilizing agent?" (a dispersion of dry diatom silica organisms or simply sawdust) A lively discussion followed among the Nobel laureates with speculations all over the place, and I recall that one was about acoustic damping, until suddenly a shout from the audience interrupted the discussion — Nobel laureate Manfred Eigen rose and claimed we were all wrong, and the true explanation of how the tricky nitroglycerine was tamed was due to the stabilizer reacting with radicals, quenching chain reactions.

We were interviewed afterwards by the editor of *Chemical and Engineering News*, Dr. Madeleine Jacobs, who turned to me as she

Photo 44.2. The Lindau Nobel Laureate Meeting in 2002. From left to right: Harry Kroto, Ahmed Zewail, George Olah, myself, Richard Ernst, and Paul Crutzen. (Taken by Madeleine Jacobs and reproduced with kind permission from Madeleine Jacobs and *Chemical and Engineering News*.)

introduced herself and said, "We have actually met before but maybe you don't recognize me with clothes on." This comment, which triggered a great deal of general amusement and not least Ahmed's, has a basis — both Madeleine and I regularly started our days with a swim and workout at the hotel pool before breakfast. Being the only users of the large pool at that time in the morning, we went through a couple of days of seeing and greeting each other at a distance, but I had obviously not recognized her "with clothes on" and swim cap off when she reappeared in the conference. Another amusing incident occurred when I was checking out from the hotel and had no cash (the hotel refused to take any cards). In line behind me was Dr. Lorie Karnath. Anxious as she had to catch a flight and possibly also feeling pity for me, she bailed me out and paid for my room and thus rescued me from a potentially embarrassing fate. In return, I later treated her to a nice meal and great wine at the Grand Hotel in Stockholm. Together with Professor Magdalena Eriksson, Lorie later became instrumental during the first stumbling, founding steps of creating the Molecular Frontiers Foundation; both of them are still highly active core members of the organization today.

A special memory I have with Ahmed was one summer afternoon when he was on his way from Lund to a spectroscopy conference in Copenhagen, and had made a detour to our summerhouse at a little fishing village by the Sound between Sweden and Denmark. He stayed for an early dinner, but as he was about to call for a taxi to take him to the hydrofoil ferry in Malmö, I suggested, "Why don't we sail you over and save you some money?" Soon we were under sail with my neighbour and very close friend, orthopedic professor Björn Persson, in his boat heading straight west. The wind was brisk but not cold and we sailed fast. After a little more than two hours, we entered the mouth of the harbour of Copenhagen, reduced sail, and moved slowly into the very heart of the continental city. Both Ahmed and I remembered the event fondly — that was the first time he ever sat on a sailing yacht; his only experience of boats was back on the Nile when he was a boy. He enjoyed every moment, especially the magical feeling as we silently approached the big city in darkness with its many lights ashore. We landed in Christianshavn, a part of Copenhagen which is a little like Venice with narrow channels and many small restaurants. It was close to midnight and we wondered if it would be possible to get anything to eat before Ahmed departed for his hotel and we returned to Sweden. Close to where we had docked the boat, we found a small pub in the basement which was closing and seeing off its last patrons. When Ahmed asked the owner whether we could have something very simple to eat, the owner was most reluctant, but Ahmed's charm quickly broke down any resistance and soon we found ourselves sitting together with him and the kitchen staff having a veritable feast. One of the waiters, who was a Portuguese, even took out his guitar and entertained us all with emotional Fado songs!

The last time I saw Ahmed was when Gunnela and I visited Pasadena in November 2015. I gave a seminar in his group on our recent discovery of a new elongated conformation of double-stranded DNA (the "sigma form"), its potential biological role, and why any form of organic life can only exist in water-rich environments. He raised many good points and also suggested a clever mechanistic complement to my explanation of the inhomogeneous conformational reorganization which we call "disproportionation." Earlier that day, he had shown me some amazing results from time-resolved single-electron microscopy, visualizing in real time the coherent travel of phonons along a set of hydrocarbon chains aligned at a surface. At the end of the chain, the wave was reflected and returned back. With a twinkle in his eye, he said, "You saw it in a femtosecond, didn't you?"

The last photograph (Photo 44.3) I have of Ahmed and myself is from my 70th birthday symposium in Gothenburg in early May 2015. As usual, he gave a captivating talk (which can be found on www.MolecularFrontiers.org) which was appreciated just as much by the professional scientific audience which included many Nobel Laureates as it was by the more than 200 high school students in attendance.

Ahmed, I am most grateful to you for all inspiration you gave me over the years and for your heartfelt friendship, the memory of which I will carry in my heart for as long as I live.

Photo 44.3. Ahmed and I at the Molecular Frontiers Symposium in May 2015. Reproduced with kind permission from the photographer, Mr. Jonas Ekberg.

Bengt Nordén is a professor of physical chemistry at Chalmers University of Technology. He is the author of 500 papers on recognition mechanisms of nucleic acids and molecular spectroscopy with polarised light, 100 encyclopedia and popular science articles, and two spectroscopy text books. He is also the former chairman of the Nobel committee for chemistry and has been elected to the Royal Swedish Academy of Sciences, Royal Physiographic Society (Lund), Royal Society of Arts and Sciences (Gothenburg), Royal Swedish Academy of Engineering Sciences, the National Academy of Sciences of Germany (Leopoldina), Swedish Academy of Engineering Sciences in Finland, Norwegian Academy of Science and Letters, Academy of Sciences and Letters of Finland (Societas Scientiarium Fennica), Academia Europaea, and Academy of Sciences for the Developing World (TWAS). His honorary degrees and fellowships of universities and societies include the United Kingdown (HonFRSC), Singapore (NTU), Australia (ANU), China (Sichuan), and India (Chemical Society). He is the recipient of the Goran Gustafsson Prize (1992), the King Abdullah University of Science and Technology Award (2008), the Arrhenius Plaque (1994), and the Arrhenius Gold Medal (2009). He is the founder of the *Molecular Frontiers Foundation*, a global organization hosted by the Royal Swedish Academy of Sciences with the objective of identifying breakthroughs in science and to stimulate young people's interests in science through the molecular sciences.

45 A Mini Voyage through Time: From *n*-Butane to Biological Macromolecules

Dmitry Shorokhov*

In the fall of 2003, I, then a fourth-year postdoctoral scientist in the physics division of the University of Augsburg in Bavarian Germany, spontaneously decided to contact Professor Zewail to see whether he would be willing to offer me a postdoctoral position in his group. Frankly, I did not expect to hear anything back from him, but to my great surprise, he responded almost instantaneously with a job offer. When I told Sanae — my wife and my guardian angel — that we were moving to a Los Angeles suburb because a Nobel laureate working there had offered me a job, the only question she asked me was: "What is the distance from that suburb to Beverly Hills?" "Well," I replied, "I don't know yet, but you will have a fair chance to figure it out." Upon my arrival at Caltech in 2004, I was overwhelmed with what I saw there: *"pretty girls, big operators, and so on."* Caltech was like Hollywood in the science universe; potentially Oscar-winning stories had been unfolding on campus in real time according to the weekly newsletter and monthly science magazine I began receiving as a newly-hired postdoctoral scholar in chemistry. What was going on in the chemistry division back then was, indeed, very exciting, but my self-confidence inevitably dwindled as I could not imagine

*Senior Staff Scientist.
Email: dikun@caltech.edu

myself delivering on par with all these brilliant Techers. For a shy and borderline autistic introvert from a relatively small, chilly, provincial Russian town, "californification" was, apparently, quite a challenge!

Some twelve and a half years later, although my memory generally does not appear to serve me particularly well, I do recall what happened during my very first encounter with Professor Zewail at his private office in 116 Noyes with amazing clarity. The office was just enormous, but right away I noticed that it was furnished rather tastefully, with great attention to detail, quiet elegance, and a bit of a personal touch (there was a nicely framed, large picture of King Tut — which had been starring in many television documentaries — hanging above a cosy sofa). Professor Zewail was on a conference call at his impressive desk; when the call was finally over, he slowly turned toward me in his massive leather armchair. Things did not develop smoothly at the very beginning and there was a certain amount of electricity in the air. Because I did not have an e-mail account at the time, Sanae contacted Professor Zewail directly to see where I was, and she was careless enough to misspell his last name. "What is your expertise?" — I recall him asking me repeatedly in his loud voice. Because I did not know yet what my true area of expertise was going to be, as a former electron and X-ray diffractionist, I ended up being assigned to a project in gas-phase ultrafast electron diffraction (UED). We had a series of warming-up meetings and, following a discussion with Professor Roberts which took place in the conference room across the hallway, I ended up doing some quantum chemistry calculations on my constantly overheating laptop.

It was pretty clear to me from day one that this was not the right way to go about it. Also, because the UED technique had been around for more than a decade — and because all the low-hanging fruit had already been harvested — UED per se did not interest me much. Last but not least, the approach was not really suited to explore conformational changes in the biological macromolecules that fold, misfold, and perform their functions in aqueous solution — and that caught my eye in late 2004. In a classical UED experimental setup, the internuclear distance distribution across a gas-phase macromolecular ensemble can, in principle, be compared with that calculated using a plausible theoretical model. However, "real-world" UED observations of order–disorder transitions in biological macromolecules are hampered by a number of serious issues such as dehydration, conformational flexibility, temperature uncertainty, finite data range, and coarse detector binning, to name just a few. Back then, there had long been the realization that a gap existed between what people ideally wanted to do on the biological frontier on the one hand, and what they were realistically capable of doing using standard experimental and theoretical techniques on the other; we needed to advance beyond molecules like *n*-butane to macromolecules like proteins and DNA/RNA. To bridge this tantalizing gap, a whole new way of thinking was required.

The breakthrough came with the realization that (non-equilibrium) conformational interconversions across the entire ensemble could be simulated *numerically* with a great deal of accuracy. Massively distributed, all-atom molecular dynamics calculations comprising hundreds of independent mechanical motion trajectories and performed in explicit aqueous environments readily provided detailed structural insights into how exactly macromolecular unfolding — and folding! — processes occur (and with the atomic-scale spatiotemporal resolution unprecedented hitherto), but carrying out such simulations would not be feasible without acquiring a state-of-the-art, professional-grade supercomputer cluster that had to be put into service in our laboratory. As a senior staff scientist overseeing network-attached storage and high-performance computing facilities at the Physical Biology Center for Ultrafast Science and Technology (UST), which was established at Caltech by the Gordon and Betty Moore Foundation under the supervision of Professor Zewail in early 2007, I have been mostly focusing on the order–disorder transitions in helical biopolymers, such as DNA/RNA and α-helical polypeptides. If there is one single characteristic of biological macromolecules, it is the presence of helices. Although the thermodynamics of their formation was well understood beginning with the 1950s, the actual dynamical pathways and transformation rates involved have remained an area of active research ever since.

In 2014, armed with powerful *4D computational microscopy* capabilities providing insights into both structural dynamics and chemical kinetics, we were able to demonstrate for the first time that zipping in polyalanine, despite its (strongest) α-helix-making propensity, gets "caught" by nonhelical contacts that block growth. These β-hairpin-like misfolded intermediates dominate the folding, leading to zipping times that are up to an order of magnitude longer than those expected based on unimpeded growth of α-helices. This observation was in perfect agreement with some puzzling experimental results and it finally explained why α-helix folding timescales strongly depend on the actual protein sequence involved. Also, the intermediate structures observed during the course of our ensemble-convergent numerical experiments have implications for protein aggregation, which is of significance in diseases such as Alzheimer's. Stimulated by the clear-cut experimental evidence of non-two-state DNA (un)folding kinetics obtained at the UST in 2007, we are currently in the process of extending the general analytical method for constructing dynamical free energy landscapes of α-helical polypeptides to a comprehensive theoretical model of chemical kinetics and structural dynamics of DNA/RNA.

Professor Zewail in 116 Noyes at Caltech (picture taken by the author, July 31, 2014).

P.S. A jammed, torn, and yellowish hard copy of Professor Roberts' e-mail to Professor Zewail dating back to the pre-historic era of the mid 2000s, which I have just been able to miraculously excavate, reads: "Pharaoh! When your guys did *n*-butane, did they see the steric effects for the $-CH_2-CH_2-$ segments? … Peon." Had I not found the text of the letter of interest back then, I would most likely not have kept it together with an (equally strongly abused) original of our marriage certificate which celebrated its 15th anniversary on 13 August 2017, and a small bunch of ancient family photographs which, I thought, had disappeared forever. Science-wise, we have come an astonishingly long way, but the line of research that begun at Caltech with *n*-butane is, pretty amazingly, still going strong.

Acknowledgements

I am indebted to Dr. Dema Faham and Dr. Milo Lin for their helpful comments.

Dmitry Shorokhov was born in Russia in 1971. He received his B.S. (1992) and M.S. (1995) in applied mathematics and physics from the Moscow Institute of Physics and Technology (the renowned "Phystech"). Thanks to unmatched flexibility of the so-called "Phystech System" largely modeled on Caltech curriculum, he was allowed to present a master's degree thesis in statistical mechanics while majoring in electrical engineering and computer science. In 1996, he was admitted to graduate studies at the University of Oslo, where for his Ph.D. in physical chemistry (2000) he studied conformationally flexible molecules, including a DNA base, by gas electron differaction. Upon receiving an Alexander von Humboldt postdoctoral fellowship, he spent four years at the Technical University of Munich, and subsequently the University of Augsburg, performing experimental and theoretical electron density distribution studies. Another spectacular turn in his career came about in 2004, when Prof. Zewail invited him to join the research team of the Laboratory for Molecular Sciences. While at Caltech, he has been overseeing the acquisition, construction, and operation of a massively-distributed computing facility as a senior postdoctoral scholar (2007), staff scientist (2007), and senior staff scientist (2008). At present, his research interests encompass computational molecular biology and high-performance computing.

46 I'm Just Thinking Aloud with You!

Milo M. Lin*

There is a photograph of the Zewail laboratory members sitting around a beach campfire in El Capitan in 2005, wherein the camera captures Ahmed Zewail's infectious intellectual joyfulness. During these moments, you knew he would transition into brainstorming mode if he said, "I'm just thinking aloud with you!" Following those words, he would take you on a journey to explore the possible approaches and potential consequences of the essential core concept behind your project. In this way, he made everyone around him feel that their particular quest was the essential question of the age. To me, he would say something like, "We cannot yet condense the multidimensional complexity of living systems into a few simple collective variables, but just imagine if we could!"

I remember him in brainstorming mode in the spring of 2006. I was ready to embark on a Ph.D. in theoretical physics, having convinced myself that string theory was the thing to do and Stanford was the place to do it. Even at that time, I felt myself drawn to questions of how biological complexity arises from simple physical interactions, yet the feeling "in the air" was that the "best" physicists were

* Graduate Research Assistant.
Email: milo.lin@utsouthwestern.edu

supposed to be trying to unify all the elementary physical forces. It was at this critical juncture that the professor invited me to join him on a stroll down the Caltech Olive Walk. His message was simple: despite the accumulation of vast amounts of biological data, we do not have a general theory of biological self-organization based on simple axioms that yields testable predictions. The need to find this theory (if such a theory even exists), he correctly intuited, was the real fuel for my scientific curiosity. He convinced me that the time was ripe to tackle this question, that Caltech was the place, and that I was the one.

During my Ph.D., although I was sometimes nervous about my meetings with him (especially when there was no progress for a while!), I always looked forward to them because I was sure to learn something about my project, about him, or about myself. These meetings had a slow, monastic focus; no matter how quickly my thoughts and speech were racing, Zewail would reset my pace, usually with exhortations to take a deep breath and have a "civilized" discussion with tea and sometimes chocolates. Every conversation had an informal formality, right down to the scholarly preamble before diving into the scientific discussion. It seemed like he was channeling the habits of scholarship from a bygone era, an era in which science was primarily an intellectual fellowship rather than a career path and scientists produced fewer, yet — to paraphrase Gauss — riper papers. It wasn't until much later that I understood how these rituals made me aware of and grateful for the privilege of a life spent satisfying my curiosity. This was a continuation of an intellectual heritage going back to the gentlemen-scholars of the Enlightenment, and even further back to the ancient Greeks. During one of our preambles, our conversation drifted to the ingenuity of the classical world. When I mentioned the Greek mathematician Eratosthenes' elegant measurement of the Earth's circumference, he firmly reminded me that Eratosthenes worked in *Egypt*, and was in fact the chief librarian of the library at Alexandria, near his hometown.

This sense of continuity with the distant past, even over eons, seemed natural to him; the convergence of cultural history and personal purpose seemed to frame his whole career, down to the central importance of light, both in his scientific legacy and personal identity. Every time I passed by the framed photograph of Einstein outside his office, I would reread the line reprinted below the photo: "For the rest of my life I want to reflect on what light is." This obsession with light even permeated his linguistic habits; instead of "showing," he was always "elucidating"; instead of being "interested," he was always being "excited" and "stimulated."

To Ahmed Zewail, the past is not dead; in a sense he believed that Alexandria's library is still standing, and so too its lighthouse — one of the seven wonders of the ancient world and a symbol of the perpetual fight against ignorance. To me, it's no coincidence that he chose to build his career at Caltech, whose sigil is the flame of discovery. He inspires me to carry the flame, not to where it is most promising, but to where it is darkest.

Photo 46.1. Zewail group retreat, El Capitan State Park, August 2005. From left to right: Ahmed Zewail, Milo Lin, De Ann Lewis, Jerry Yang, Ramesh Srinivasan (standing), Chaozhi Wan, D. Hern Paik, Sang Tae Park, Jon Feenstra (foreground).

Milo Lin was born near Wenling, Zhejiang Province, China in 1984. He immigrated to the United States at age 7, growing up in Socorro and Los Alamos, New Mexico. After obtaining a B.S. and Ph.D. in physics from Caltech and spending three years as a fellow at the Miller Institute for Basic Research in Science at the University of California, Berkeley, in 2015 he became an assistant professor in the Green Center for Systems Biology at the University of Texas Southwestern Medical Center. He has joint appointments in the Department of Biophysics and Center for Alzheimer's and Neurodegenerative Diseases. His research goal is to understand how biological systems find functional solutions in the vast space of possible solutions on physiological and evolutionary timescales. He currently lives with his wife, Megan Tom, and their dog, Barrington, in Dallas, Texas.

47 A Recall of My Stay with Dr. Zewail at Caltech

Aiguo Wu*

It has been over ten years since I left Caltech as a postdoctoral scholar in 2006, but many things remain vividly memorable. I will enumerate several examples to memorialize Dr. Zewail.

I arrived in Pasadena with my wife on 25 July 2005. Dr. Liang Zhao, a former postdoctoral scholar in the AZ group, fetched us from Los Angeles Airport to the campus by car. In the afternoon of 27 July, Dr. Zewail asked his secretary, Ms. De Ann Lewis, to call me and asked me to go to his office to discuss my potential research projects at Caltech. Dr. Zewail got me to present and discuss my previous work, including why I chose to do what I did, what my motivations were, what had been accomplished so far, and what kinds of scientific problems need to be resolved, amongst other things. After the discussion, Dr. Zewail put me in the ultrafast biological group together with Dr. Chaozhi Wan and Dr. Hairong Ma to do some research work. I learned a lot more about science just from that face-to-face discussion with him and also realized that I had a long way to go in my future research.

The second memorable event I want to recollect is the Thanksgiving Day party in 2005. On that day, Dr. Zewail invited over ten group

* Postdoctoral Research Fellow.
Email: aiguo@nimte.ac.cn; Homepage: http://wuaiguo.nimte.ac.cn/

members and their families to go to his house in a suburb. Everybody arrived around 5.00 p.m. While De Ann distributed the roasted turkey meat and food to each person, Dr. Zewail prepared a video about his Nobel Prize-related issues for us to watch. On that wonderful evening, we enjoyed the turkey meat and other delicious food as we watched the video. At the same time, he stood by the side of the video screen and spoke to us of more details whenever he felt that more elaboration was needed. He also mentioned that before he finally won the Nobel Prize, he would escape from the Caltech community and Pasadena every October because many people would think that he would get the prize that year, but year after year before 1999 he failed to get it. The pressure was certainly on and mounting as to whether he could win the Nobel Prize every October! On that Thanksgiving evening, he showed us another side of the Nobel glory using his sense of humor.

The third thing I want to recall is a group retreat that took place in Santa Barbara in August 2005. We had spent two days in Santa Barbara. On the first afternoon, we discussed mostly science-related issues person by person. After supper, many group members sat together around a bonfire and discussed things that were more outside of science. For example, Dr. Zewail talked about the political systems of the United States, Egypt, and China. He thought that the United States runs a democratic system on the surface but is actually controlled by capitalists. Egypt was learning and adopting modern Western political systems such as that of the United States, but there were a lot of problems which made the Egyptian government unstable. On the contrary, China runs a centralized system (or even, to the West, a totalitarian system) which might make it easier to face great challenges together. Most of us did not realize until then that he had such deep thoughts about social and political issues around the world. The developments in different countries caused by globalization have shown that his ideas were quite spot on. In my opinion, he was not only a great scientist but also a great political thinker, and this side of him was reflected in his political efforts to his homeland, Egypt.

The last thing I want to mention is a small gesture from him which meant a lot to me. In 2007, after I completed my work contract at

Caltech, I left Pasadena and moved to the Feinberg School of Medicine at Northwestern University in Chicago. One day, I received a yellow parcel from Caltech, and I was very touched when I opened it and saw the contents. There were six hard copies of a paper published in *Proceedings of the National Academy of Sciences* where I was listed as a co-author. I then recalled that, during a short chat with him in his office just before I left Caltech, he had told me that I would be listed as a co-author on that project which was highly relevant to my research area. Such a kind person!

When I returned to China in 2009 and got a permanent research position at the Chinese Academy of Sciences, I sent an email to tell him about the good news. He asked De Ann to send me a greeting email on behalf of him, conveying his joy to learn of my new appointment.

Thus, these stories show that Dr. Zewail is not only a great scientist but also a deep thinker and a very kind man! We will remember him from the bottom of our hearts.

Aiguo Wu received his bachelor with honors from Nanchang Univeristy, China in 1998; and received his Ph.D. from Changchun Institute of Applied Chemistry, Chinese Academy of Sciences (CAS) supervised by Prof. Erkang Wang and Prof. Zhuang Li in China in 2003. He was at the University of Marburg (Prof. Norbert Hampp's group) in Germany during 2004–2005; Caltech (Prof. Ahmed Zewail's group) in USA during 2005–2006; and Feinberg School of Medicine, Northwestern University (Prof. Gayle Woloschak's group) in USA during 2006–2009. In 2009, he joined Ningbo Institute of Materials Technology & Engineering, CAS as a full professor and Principal Investigator. Prof. Wu has published over 130 papers in PNAS, *Chem Soc. Rev., Adv. Mater., ACS Nano,* and *Biomaterials,* etc., one book in English and three book chapters, and has been awarded 43 invention patents. His lab focuses on using nanoprobes for early diagnosis and therapy of diseases, such as cancer and applications in environmental sciences, etc.

48 My Personal Memory of Dr. Zewail and his Femtoland Frozen in Time

Oh-Hoon Kwon*

The loss of Dr. Zewail came to me so suddenly. When I was informed about the news from Pasadena almost a year ago, I simply regarded it as a bad joke. It was only until recently that I could finally come to terms with the loss. The incident drove me to recall the precious and joyful moments I had with him for eight years in "Femtoland." These include his impact on a broad range of scientific (and non-scientific) communities. The reminiscing and rumination of these great memories finally helped relieve me from his absence.

This memoir of Dr. Zewail, which was strongly encouraged by Abderrazzak, Dongping, and Spencer, reflects only some of my recollections and brings me all the way back to the crystal-clear memory of me greeting him at his office on my first day in Pasadena. On October 1, 2005, I had been so nervous not to give him a lousy impression at the first meeting because I had been told back in Korea that he held very high expectations of his group members. I quickly learned that he loved cheering young minds on and was full of humor, thus making me and many others feel very welcomed. Among his many witticisms, for example, when he wanted to find me quickly to discuss new results or an inspiring idea that had just popped up in his mind, he would call

---

* Postdoctoral Research Fellow and Senior Scientist.
Email: femtokid@gmail.com OR ohkwon@unist.ac.kr

me and ask over the phone, "Oh-Hoon, could you come upstairs in a femtosecond?" To be sure, I could never make it in a femtosecond, but I would certainly try my best not to lag too long from "time zero" of the call so that I could catch him while he was still in the "excited state."

He was always bursting with energy. I remember him regularly coming to the office very early in the morning, second only to Jon Feenstra, a Ph.D. student in the same group who used to get in before 4.00 a.m.; I was usually the third showing up around 7.30 a.m. One day, Dr. Zewail told me that his brain was divided into twenty sections, each reserved for the research of each group member. But he was no exception to being part of the same universe as we were and his time passed at the same rate as us. Yet, he still always managed his commitment to research, mentorship, and service without any wastage of precious time in remarkable ways. Truly, he lived his life in femtosecond-time resolution. Because it was always hard to arrange a meeting with him, whenever I got a chance to sit with him, I tried to catch his interest with new results and convince him of their significance. In those meetings, Dr. Zewail never wasted any moment; he would get to the point quickly and give very insightful feedback.

Not too long after joining his research group, I realized every single word and gesture of his was meaningful and intended to convey a message to young, curious individuals. One day, he found me and gave a metallic bookmark with a famous quote by Winston Churchill: "Never, Never, Never Quit." Back at my office in the subbasement of Noyes, I descended into a bottomless mental abyss trying to decipher the meaning of that gift — did he want me to stay around much longer than I anticipated? On another day, when he returned from a business trip to Australia, I was given a big wooden boomerang as a souvenir. What was this supposed to mean this time? Was he telling me that no matter how hard I try to find a new job elsewhere, I would end up returning back? In retrospect, these were really meant to be thoughtful encouragements; thanks to his warm mentorship, passionate guidance in pushing the frontiers of science and technology, and kind sharing of his vision with me, I am now fortunate to be able to extend his legacy in Korea, returning to the field of fancy ultrafast electron microscopy from spectroscopy!

One day in 2010, I made a really big mistake. A meeting had been arranged between Dr. Zewail and me and our plan was to start writing a manuscript from early in the morning. Against my better judgment, I was still drinking wine the night before until 4.00 a.m. at my friend's place somewhere on Mar Vista and could catch only two hours of sleep. I bet when I stepped into his office, he must have noticed that I was heavily drunk. I barely survived writing that paper with him from 7.30 a.m. to noon when I was finally released from the office for lunch (or more precisely a nap to me) to the tune of De Ann's worried query, "Oh-Hoon, are you OK?" Dr. Zewail was kind enough not to mention anything about my physical state during the meeting; I guess I was ultimately still punctual to the minute and somehow managed to fulfill my duty despite being drunk. Since then, however, I got bestowed the title of "party boy" from Dr. Zewail until I left the group in 2013.

On a sunny day, as it always is in Pasadena, about a week or two after our manuscript on 4D electron tomography was accepted for publication, I was invited to Dr. Zewail's home. We reminisced about the excitement of the work and discussed possible future projects. After smoking some cigars for about an hour, Dr. Zewail shot me the famous question: "Is the glass half full or half empty to you?" Although that expression is well known, that was the first time I heard it. He wanted to tell me that the journey to seek unprecedented knowledge is always challenging, so one should be optimistic and enjoy the path, otherwise one is prone to be discouraged and fail. Now back in Korea, whenever challenges stand in my way, I remember that moment with him, gaze at his gifts, and repeat that quote which I have kept deep in my heart.

A popular Asian proverb says, "Tigers die and leave their skins. People die and leave their names." To me, Dr. Zewail left not only his name through his great scientific contributions to the world but also through the people who inherit his philosophy. His legacy will carry on through me in Ulsan as well as others all around the globe for the advancement of science and humanity.

Dr. Kwon received his B.S. degree from Seoul National University, Korea in 1998. To fulfill his military service, he stayed in Korea and completed his Ph.D. studies on picosecond-resolved spectroscopy in 2004 at the same university. Following his postdoctoral training at the California Institute of Technology under the supervision of the 1999 Nobel laureate in Chemistry, Dr. Ahmed H. Zewail, from 2005 to 2010, he served as a senior scientist in the same institute. Dr. Kwon has co-authored with Dr. Zewail 20 research papers and earned one patent on the development of 4D electron tomography. As a member of a collaborative team, Dr. Kwon developed the second-generation ultrafast electron microscope and also reshaped room 036 of Noyes with the development of fluorescence upconversion and time-correlated single photon counting spectrometers. In May 2013, he returned to Korea after accepting a faculty position in the Department of Chemistry at the Ulsan National Institute of Science and Technology. One of his current interests is the development of a robust ultrafast electron microscope and the expansion of its scope for the study of reversible and irreversible phenomena of hard and soft matter. His recent recognitions include the Distinguished Lectureship Award from the Chemical Society of Japan, which was awarded in 2015.

49 Pizza, Mandolin, Pyramids, Carpets, and Science

Fabrizio Carbone*

Many have said that if Dr. Zewail lived during the time of ancient Egypt, he would have been a Pharaoh. I think that he would have rather been the man who told the Pharaoh what to do.

I met Professor Zewail in 2006. I was actually with my wife on our honeymoon which we combined with my job interviews in California. I remember having a few different options at that time, but I was truly fascinated by Caltech and the experiments that were being done in AZ's laboratory. He was very kind and polite with me, but I could also see that he was cautiously scrutinizing me. I came from a rather different field and thus he could not identify me very well. I understood immediately that there was no joking around there. He was very meticulous in ensuring that only the best people could join his group. So, the only way in for me was to win a fellowship to support my postdoctoral work there. I proposed this possibility to him and he obviously accepted it. Among all the places that I visited and that offered support for my application to such a fellowship, his invitation letter was the weakest and the least committing. I eventually still decided to choose Caltech as my host institution. A few months

* Postdoctoral Research Fellow.
Email: fabrizio.carbone@epfl.ch

before I expected to know the decision outcome of the fellowship and start my postdoctoral work, AZ contacted me to ask when I was planning on going there. I was surprised by his sudden interest in having me. He told me that he wanted me to overlap with one of his researchers who was about to leave. I explained that I was waiting for my fellowship to be approved and he asked me if he could do anything to speed things up. The fellowship decision, which turned out to be positive, came right during this e-mail exchange with him; it was clear that he could and would just go and get what he needed when he needed it.

When I joined Caltech, I remember vividly that in the first month, AZ often swung by the office to check on what I was doing and once even directly asked me if "I was wasting my time." I wasn't worried by this; in fact I was simply too excited to be able to finally put my hands on the equipment in that laboratory. It was an amazing period being with outstanding people who were working in the ultrafast electron crystallography laboratory, and when I got my time-window to use the machine, I was lucky that the project I had in mind for my fellowship worked at the first attempt. Thanks to the collaboration with the other people in the group who were extraordinarily talented, as I expected, I obtained fantastic data within a month of my arrival. I started discussing these results with AZ and from then on things kept being better and better. Later, he admitted that at the beginning he thought I was just a guy looking for a nice holiday in California. On my side, I was expecting that a Nobel laureate like him directing a large laboratory would mostly be a manager by then. That couldn't have been any farther from the actual reality. The most pleasant surprise for me was that he was just a person with an instinctive curiosity and a genuine passion for science. If you showed him a new experiment, a new effect, or some new data, you would capture his attention immediately and he wouldn't let you live a normal life until a paper was published about it.

I remember that while we were finalizing an article for *Physical Review Letters*, AZ had to take a long trip to Asia and back to Los Angeles. After a few punishing 10-hour-long connecting flights, he got stuck on the airplane at Los Angeles Airport because the computers at the arrivals went down. Altogether, he had been travelling for about 40 hours in a row. Once he cleared the airport, he went straight to the laboratory, looked for me (it was around 6.00 p.m.), and said, "Fabrizio, I had a nightmare travel, I am very tired. Do you mind if we discuss your paper at my place in my garden rather than here so that I can relax a bit and smoke a cigar?"

AZ had to explain everything — he couldn't accept it if something appeared to defy his understanding of nature. The other important aspect of his scientific personality that I soon came to understand was that he was very jealous of his ideas. He wanted to be able to voice his ideas without being biased by what other "experts" would say; my original attitude in the face of a new effect was to try and discuss it with an expert first. AZ instead would want to give his explanation first before letting other experts discuss the idea which was now contextualized by his explanation. I later understood the merits of this method, which is an excellent way for a researcher to make an impact and to most importantly advance a field. Even when an idea turns out to be imprecise, to disprove it experts have to challenge it and push the boundaries of what they already know. That is when new science emerges. I have seen this happen several times while working with AZ and reading his papers.

He believed that there was more merit in an interesting or exciting idea that was merely plausible than in a solid but boring explanation. Oh how I loved that attitude! In my view, this was the best way to challenge dogmas and to stimulate discussion and interest, which was something he did in his entire career. Obviously, it didn't hurt that besides being interesting and innovative, his ideas were often right.

On the debate between whether scientists should strive to be creative and innovative (AZ's stance) or strive instead to be always right, I would say that the choice is a matter of personal taste.

During my two-year-long stay at Caltech, I had many memorable moments not just at the scientific level but also at the human level. AZ's sense of humor was one of a kind and it did make for the following great moments of fun.

The *Condensed Matter* arXiv

After several months of struggle, we eventually got an article on superconductivity published. I went to AZ's office and suggested uploading it on the *Condensed Matter* arXiv as well, an online platform where preprints on non-open access papers can be shared for scholars to peruse and discuss. Dr. Zewail reluctantly agreed, but I could tell that he was not convinced by the value of such an initiative.

The day after I uploaded the manuscript, he called me to his office. He was all excited because he received several emails from people who read the paper on the arXiv and had questions or comments about it. So he said to me, "This arXiv thing really works! People read it apparently, I am surprised. So, what do I do if I want to upload a paper there?"

I replied, "Well, to upload articles on the arXiv, you need to be an endorsed author. For example, I could write an email to the administrators of the webpage saying that I met a promising young chemist who would like to contribute something to the condensed matter physics community."

A few seconds elapsed while he was considering how serious I was and the baffled look on his face prevented me from holding back a laugh.

The Earthquake

One day, I was working in the laboratory with my colleague, Dr. Petros Samartzis. We were sitting beside a 100-litre tank filled with liquid helium. At some point, everything suddenly started shaking a lot. "What's going on?" I asked. "It's the earthquake! Run!" Petros screamed. Without thinking, I followed his advice and ran like Usain Bolt out of the laboratory and down the corridor in the subbasement. California's weather is generally pleasant and it wasn't uncommon to wear flip-flops even to work. While running, I lost my flip-flops somewhere along the corridor and rushed out of the building barefooted. Once outside, I met Dr. Zewail. "What are you doing without shoes?" he asked me. "Well, we felt the earthquake in the lab, Petros told me to run and I just did it. I was wearing flip-flops and I lost them in the corridor." Meanwhile, my colleagues came out of the building laughing as they found a pair of abandoned flip-flops along the corridor. Dr. Zewail explained that they belonged to "Fabrizio the lionheart." I told him, "Indeed, I am not nor do I want to be a very brave man; do you still want me as a postdoc?" He quipped, "You were brave enough to come to work for me, so I'm fine. But don't wear flip-flops in the lab!"

Farewell

Before moving back to Europe, I wanted to leave something behind for Dr. Zewail, a gift of sorts that could remind him of our several discussions about Mediterranean culture. I chose three books from Umberto Eco and I wrote the following message to him:

> "Alessandria in Italy, where I come from, was founded as a stronghold in defense of a league of small towns against the German emperor Frederick Barbarossa. It was named in honor to the pope Alexander, proud enemy of the emperor.
>
> During the last siege of the imperial army, a peasant named Gagliaudo had the idea to set free a cow overfed with grain. When the troops, after slaughtering the cow, saw how well it was fed, concluded that the city could hold fast for a long time and withdrew.
>
> In Egypt, the city of Alexandria was founded by Alexander the great, and it is said that its borders were traced with grain, an omen of future wealth.
>
> Apparently, grain, the ancient symbol of prosperity and well being, has blessed both our origins. Consistently with such tradition, I leave Caltech with a wealth of scientific progresses that I feel we have made together.
>
> I wish to thank you in writing, since 'verba volant scripta manent', for the time you dedicated to me. As I said many times now, working in your group has been a dream come true. I add to these words a present, which is irrelevant in its material value, but of great intellectual and cultural significance.
>
> These three books by the most celebrated contemporary Italian writer, Umberto Eco, professor at the University of Bologna, are fine tales of historical episodes that marked our beloved Mediterranean culture. The ability of Eco in nesting anecdotes with true historical facts, and immortal legends, makes the reader see and think with the

eyes of a middle ages man, with his rudimentary scientific knowledge and superstitions. Characteristics that look silly at the eye of contemporary men, but that at closer look reveal the path of human intellect toward the modern knowledge, and our beloved science.

I hope you can enjoy these readings.

Sincerely,

Fabrizio"

This was a private message from me to Dr. Zewail. Many years have passed since and I do not mind sharing this now.

I had the honor of having Professor Zewail at my laboratory during his visit to Switzerland in 2013 (Photo 49.1). He joked that he wanted to apply to be a postdoctoral researcher under me since my laboratory was brand new; I told him that I had to think about it because I saw him typing on the keyboard with one finger sticking out of his fist once and I couldn't afford a secretary for my postdoctoral researchers. That was another great laugh we shared and sadly the last one.

Photo 49.1. At the École Polytechnique Fédérale de Lausanne with myself, Professor Chergui, Dr. Zewail, and Professor Dantus.

50 Ahmed Zewail

William F. Tivol*

I worked with Ahmed for only a short time from 2008 to 2009, but that was enough to leave an indelible impression on me. The problem I was working on was the most technically challenging one of my career, and this kind of bold research was a hallmark of Ahmed's work. He had built his reputation in ultrafast electron diffraction and decided that ultrafast electron imaging was his next field. I met him earlier in the year when he invited me and Professor Grant Jensen to discuss our experience with electron microscopes with him. I was particularly impressed that Ahmed's primary concern was our experience with the company's record servicing their products.

I started working with him on ultrafast imaging of biological materials, which entailed looking at systems that were more susceptible to radiation damage than the materials he had done most of his work with. Furthermore, imaging such materials was by phase contrast, which was easiest with a coherent electron beam — the opposite of what can be produced with an ultrafast instrument. In my first week, he had me present the theory of phase contrast electron imaging to his group. He asked many penetrating questions during the presentation, not only to satisfy his own curiosity, but also to ensure that the rest of his group could follow what I was saying.

---

* Senior Scientist.
Email: wtivo@sbcglobal.net OR wftivol@ibl.gov

Although progress was slow — I was only able to show that I could achieve roughly 1 nm resolution — he was always encouraging and upbeat about the project. I was able to propose only a few possibilities that might be achieved, and the one that excited him the most was to examine stretch receptors by attaching them across holes in a carbon substrate, exciting them by heating the substrate and causing the holes to expand, and then imaging the individual molecules as they changed conformation in response to being stretched. Even this short description highlights some of the difficulties, such as imaging individual protein molecules, imaging in a liquid water environment which is necessary to preserve biological function, attaching the protein to particular places on the substrate, and, of course, isolating the protein in the first place.

It was a great pleasure to be able to work in a research environment where such high levels of enthusiasm for such challenging projects were the standard. All my discussions with others who have worked with Ahmed have revealed the same sense of enjoyment for working not only at the edge of the possible, but seemingly even beyond the edge.

William F. Tivol received his B.S. degree in physics at Caltech in 1962 and his Ph.D., also in physics, at the University of California at Berkeley. He had many postdoctoral positions in biochemistry before taking a position as the physicist-in-charge of a high-voltage (1.2 MV) electron microscope at the Wadsworth Center in Albany, New York. After over 20 years there, Dr. Tivol took a position as facility manager of the cryo-electron microscopy laboratory at Caltech. After six years, he worked with Dr. Ahmed Zewail on biological imaging with ultrafast electron microscopy. Dr. Tivol currently works at the Lawrence Berkeley National Laboratory.

51 The Fast and the Slow in Life

Wenxi Liang*

In the blink of an eye, Dr. Zewail has already left us for more than half a year. There is an old Chinese saying that time flees like a flying arrow. In the studies we did together with Dr. Zewail, we handled events that occur really fast, measured by one million-millionth of a second, and we described them as "ultrafast." His passing was unexpectedly fast in a life time, which is counted by months and years in contrast. I spent more than six years in Pasadena working with Dr. Zewail. This was indeed quite a long time given my prime age, but I did not expect that just only a year later — that fast — I would learn of his passing.

My life at Caltech could be described as a rhythm of days fast and slow — collecting data regardless whether it was day or night, handling the results, and preparing new experiments for the next time we could use the instruments. This was more or less like the interactions we had with Dr. Zewail. He was loaded with many duties, some of which even came from the White House, so sometimes we did not get to see him for several weeks. "Is he upstairs?" was a common question among us in the laboratory. We usually had to fix an appointment in order to discuss things with him, and we often did not expect that any request would lead to a quick or successful arrangement. But once a significant progress was made in our

* Postdoctoral Research Fellow.
Email: wxliang@hust.edu.cn

research, his attention would be drawn and highly focused on it. He would spend most of his time on that case for several days in a row, working on the discussion of the data, the scientific story behind the results, and even the words and punctuations used in the manuscript write-up. Life at this point would turn into a marathon lasting days or weeks together with Dr. Zewail.

I remember working on a particular paper once with him. The writing started in the morning in Dr. Zewail's office and we had lunch at his desk. We did not even walk out of the office to get the meals; they were delivered to us by De Ann, our group administrative officer. Such was our routine for a few days. As the writing of the paper came to a close after lunch on the last day, I did a good stretch and said, "Oh, finally this marathon is coming to an end!" Dr. Zewail was apparently quite satisfied with this description of our work and said with a smile, "This is just like a marathon, huh?" After all that hard work, I certainly felt tired even though it was still early in the day. But that was when it struck me — on my side of the desk sat my colleague and I, both aged 30, but on the other side sat an old man approaching 70, and yet he was still going strong.

Dr. Zewail's research did not slow down even on the weekends or during episodes of illness. I had been called several times to his home to work on experimental results or papers. Compared to the office, working at his home made me even more nervous because, being a cautious person, I felt the added need to take care of my bearings on top of my work. Mrs. Zewail always served us sweet, delicious pies. I recall that I did not mind working on an empty stomach and actually preferred not having pie, but Mrs. Zewail put some pies and tea on the coffee table as we discussed the experimental results in the living room. First, Dr. Zewail said, with gentle, formal courtesy, "Thank you, Madam." (Later I noticed that he did the same each time he received things from Mrs. Zewail.) Then he invited us to try the pie. I nervously cut and carefully ate the pie, fully worried that the crumbs would land on the carpet. I felt relieved only after the empty plate was finally laid safely on the table without incident. My cautious nature transformed the experience of eating a delicious pie into one of hard work and torment.

Later, I learned to let go of this cautiousness — the pies provided by Mrs. Zewail were truly a refreshment! There was one particular afternoon where we worked at Dr. Zewail's home again, but it wasn't for anything really urgent, just the slight revision of a manuscript. The work was done in a short time and then the delicious pies came again. After the tea break, we sat back in our chairs, chatting casually about the cultural differences between the United States and our homelands, the hot current affairs stories in the Middle East and Europe, and so on. On that afternoon, the sunshine of Southern California was tender, the warm wind of spring was soft, and lounging in the yard was so relaxing. The time went by slowly like a brook on the plain.

Although Dr. Zewail was serious and careful at work, he also liked making jokes to put us at ease. As one of his typical ways to remind us to work hard, he would say, "Please do not enjoy too much the sunshine and beach in California." Whenever we approached him, often because we were going to propose something, he would say with a smile, "What can I do for you, Sir?" Of course, whether we could get his approval or not was another story. Dr. Zewail enjoyed employing trickery in his words. There was once my colleague Sascha and I reported to him on the progress of upgrading the ultrafast electron crystallography instrument. Based on the results at that point, the progress needed more tests of heavy loads, and we proposed a schedule for these tests. After discussing the details, Dr. Zewail proposed a shorter time frame, as he always would, and then he looked at us and said with his signature smile, "Gentlemen, same things, shorter time; do you think you can make it?" Sascha thought for a moment and replied with a shrug, "Why not, one always can be better." Dr. Zewail turned to me and asked, "Wenxi, do you agree with him?" This was an apparent trick. I spread my hands and said, "Well, at least I learned a nice English sentence today; one always can be better."

Dr. Zewail had visited some Chinese academic institutes and was an old friend to the Chinese scientific community. Coincidentally, he visited the Institute of Physics at the Chinese Academy of Sciences in 1999, which was where I would later receive my Ph.D.; in that same year he also visited the Huazhong University of Science and Technology where I am now a professor. Considering the impossibly large size of China and the sheer number of universities and institutes there, it's hard not to chalk this coincidence down to fate. One afternoon in 2010, I was called to his office room. Once I presented myself there, he showed me a diploma which certified, in Chinese, that he was awarded membership to the Chinese Academy of Sciences. He said, "I am now a member of this community. Once Dr. Lu called me to inform me that I was elected, they mailed me this. Now I have the responsibility of being a member there." There was no further purpose for this conversation besides Dr. Zewail's desire to share with me his pleasure of being awarded this membership from my home country.

Dr. Lu was the director of the Chinese Academy of Sciences at that time, and Dr. Zewail liked to joke with me using Dr. Lu's name and authority. Shortly after I joined Dr. Zewail's group, I reported some negative experimental results to him with a depressed look on my face as I felt bad. After a few discussions, he said, "Wenxi, don't worry. Just keep working hard. Once you survive in my group, I will call Dr. Lu to find you a job." Later when I had a run of good results and co-authored a few papers with him, he joked and said, "Wenxi, don't think that you can escape from me and hide in China. I will call Dr. Lu to let him know you are not good enough; then nobody will hire you." One day in the spring of 2015 when I had already scheduled my departure, we had a pleasant conversation about my future work. Even right at the very end, Dr. Zewail still did not relent on the jokes and said, "Wenxi, you should work hard. Otherwise I will call Dr. Lu to fire you." I laughed and corrected him, "Dr. Lu has retired from the director position due to his age. Now the director of Chinese Academy of Sciences is Dr. Bai." To which Dr. Zewail simply kept up with his joke and said, "Then I will call Dr. Bai!" We both laughed (Photo 51.1). Time had elapsed so quickly, even if more than six years had gone by. I have a new director for the Academy and Dr. Zewail had already fought against his illness.

He spent his life chasing the fast, and his mark on the history of science will be long and significant. Now I am following the path he

Photo 51.1. A happy ending to my postdoctoral scholarship with Dr. Zewail. On June 10, 2015, the day before I left Caltech, I took several pictures with Dr. Zewail in his office. At that point he looked great as he recovered from his illness. He was very pleased to see me all dressed up as we normally dressed casually in the laboratories. After the time of this photograph, I received again his wishes, and sadly for the last time.

paved, trying to boost my pace as the memory and feeling of being carried on his giant shoulders slowly fade away.

Acknowledgments

I wish to thank Dr. Sascha Schäfer and Dr. Aycan Yurtsever for the lives we shared at Caltech which made, and are still making, my life more enjoyable.

Dr. Wenxi Liang received his bachelor diploma from the department of physics at Tsinghua University, Beijing, China, in 1998. He received his master degree in 2001 from the Graduate University of the Chinese Academy of Sciences, Beijing. He received his Ph.D. degree in January 2009 from the Institute of Physics at the Chinese Academy of Sciences under the supervision of Dr. Jie Zhang. His doctoral thesis covered the development of the first ultrafast electron diffraction facility built in China. From February 2009 to June 2015, Dr. Wenxi Liang took up a postdoctoral position in Dr. A. H. Zewail's group at Caltech, Pasadena. During this time, he worked in the ultrafast electron crystallography laboratory. Starting from June 2015, Dr. Wenxi Liang was appointed with a professorship at Wuhan National Laboratory for Optoelectronics at the Huazhong University of Science and Technology, Wuhan, China. His scientific interests focus on the development of advanced ultrafast electron-probe techniques, femtosecond laser-induced structural dynamics of thin films, surfaces, and adsorbates, and materials with novel electronic properties.

52 A Polar Star in Science

Giovanni Maria Vanacore*

The first time I met AZ was in his office. He was seated behind his large desk; I was sitting in a chair in front of him with my shoulders slightly curved and his book in my hands, and I was quite intimidated from being in the presence of a Nobel laureate. It had been a week since I arrived at Caltech and he had just returned from a long trip in Europe. During his absence, I had the chance to informally meet the people in the group and visit the laboratories while studying as much as I could for my new scientific adventure. Many were the ideas running around in my head as were the experiments I wanted to try, creating a strange state of mind where my excitement was mixed with frustration from uncertainties about the future. With that feeling in me, I went to meet him, and it turned out to be *illuminating*. Thinking back to that moment, all the other meetings we had in the course of my four years of working with him evolved more or less in the same exact way.

At first, he would usually crack a few jokes to lighten the mood and break the ice. Being at the receiving end of a joke was always the sign of a good relationship with him. That was his way of connecting with anyone on a personal, and not only scientific, level. With me, the main theme was my compelling need for a good coffee

* Postdoctrol Research Fellow.
Email: giovanni.vanacore@epfl.ch

because... you know, I am Italian! Only after a humourous exchange would we start discussing about science. He would briefly explain his vision and, in few sentences, indicate the objectives to pursue. His ability to see clearly which paths to take was always his most remarkable quality. Every time I felt lost or unsure about what to do next, I knew I could always count on him to give me the direction I needed. In the end, we would agree on the next practical steps to take and then, while leaning back on his chair, he would slightly raise his chin and gently cast his eyes upward, finishing with a final encouragement: "...and just imagine if we would be able to see it!" Those words contained the excitement of the promise of a new scientific journey mixed with the curiosity of a new discovery. Those words were the inspirational fuel for every late night and weekend spent in the laboratory. Those words drove the anticipation for the moment when we would finally see something that, only just a while before, was merely a thought.

The last email I received from him was on June 27, 2016, a few months after I left Caltech to go back to Europe to carry on my research. It was not about work or science. Instead, it was about soccer! The Italian national team had just defeated Spain in the European Cup competition and AZ wrote me the following few words: "Dear Giovanni, congratulations on the 2-0 results, even though I have soft spots for both countries! Best regards, AZ." I like to think of that text as the moment our relationship transcended the pure professor-postdoctoral fellow dynamic and started entering a new dimension, that of a true friendship.

I worked with AZ for more than four years and it was nothing short of a great journey. His scientific acumen, his clear vision for the future, and his ability to be a true leader, someone who led by example, are the teachings that I will always treasure. On August 2, 2016, the world lost a great mind, and I have lost my polar star.

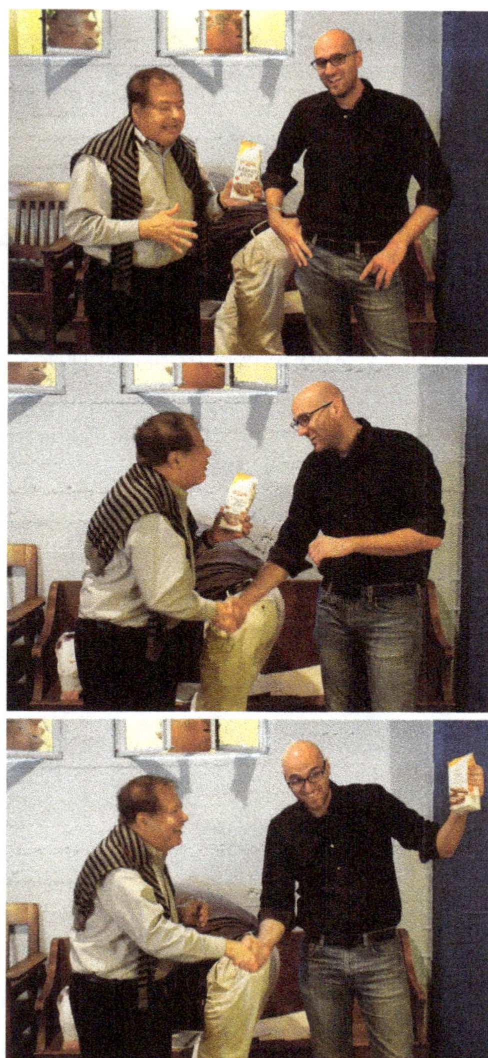

Photo 52.1. These pictures were taken on November 26, 2014 at the Atheneaum in Caltech during the annual Zewail-group lunch and portray a funny moment in the life of the group. In that occasion, one of the group members (J. Tang) organised a series of entertaining games during the meal, in which AZ participated with enthusiasm. In this moment, with a speech half-serious/half-joking, AZ was awarding Giovanni Maria Vanacore with one of the special prizes (a pack of cookies!) for having realized the highest score in one of those games.

G. M. Vanacore received his bachelor and master of science in physics engineering from Politecnico di Milano (Italy) and his Ph.D. in physics under the co-tutorship of Politecnico di Milano, École Polytechnique X (France), and the French Atomic Energy Commission (CEA). He was a postdoctoral fellow in the group of Professor A. H. Zewail at the California Institute of Technology (U.S.A.) and he is currently working as a scientist at the Institute of Physics at the École Polytechnique Fédérale de Lausanne in Switzerland. His research is focused on ultrafast phenomena in condensed matter with a special interest in the fundamental interactions among electrons, photons, phonons, and plasmons in low-dimensional nanoscale materials.

53 Endless Passion for Science: Witnessing the Last Four Years of Dr. Zewail

Jianbo Hu*

My first meeting with Dr. Zewail was in Lausanne in July 2012 when we both participated in the Ultrafast Phenomena conference. Of course, he was an invited speaker while I was still a Ph.D. student at the Tokyo Institute of Technology in Japan. But that interaction with him can be traced back to an earlier time when I applied for a postdoctoral position in his group. He had a phone interview with me and then offered me a position before hanging up. I was too excited to sleep that night; everything felt like a dream. The idea of actually working with a legendary scientist like him was too surreal. So when I talked to Dr. Zewail in person during the coffee break of that conference, it felt like my dream was becoming a reality. At the conference, I was fascinated by the exciting research he presented. Even at a distance from the speaker's platform, I could feel his passion for science (it was a really big conference hall by the way). I started to look forward to my new adventure in a different field and in a different country.

At the end of September 2012, I moved from one "Tech," Tokyo Tech, to another, Caltech, and began my journey together with Dr. Zewail through the wonderland of science. I was assigned to the

* Postdoctoral Research Fellow.
Email: jianbo.hu@hotmail.com

ultrafast electron crystallography team in collaboration with two senior members, Wenxi Liang and Giovanni Vanacore. From then on, I was lucky to have had the chance to witness Dr. Zewail's final few years of his life. It was a really precious lesson for me, not only in terms of science but also in terms of attitude to life. I had the opportunity to learn not only from his vision and passion for science, but also from his strong work ethic and will to rise against any challenge.

Dr. Zewail's health problems started at the beginning of 2013 and he received cancer treatments twice in 2013 and 2016. During this period, he never stopped working. After the first chemotherapy treatment, he lost his hair and his trademark moustache and was too weak to go to the office. So he scheduled to meet each research team at his home to keep up to date with the status of each laboratory and the progress of each project. These meetings made him exhausted even if they did not last long. As he gradually recovered from the treatment, Dr. Zewail devoted more and more time back to work. He would arrive at his office early in the morning and work till late in the afternoon. He was also present in his office on weekends and public holidays. Occasionally, we received emails from him at midnight or early in the morning. He seemed restless, working all the time. Once, I had an opportunity to observe closely how he worked. One Sunday, he brought Giovanni and me to the Coffee Bean on Lake Avenue and worked on our manuscript for hours. When he was working, he was extremely focused and dedicated. You would not even realise that he was pushing 70 and had just recovered from a life-threatening disease. Needless to say, he was still one of the hardest-working members of his group during that period.

Dr. Zewail had an ambitious plan to "push the frontier of science forward" (the exact words he loved to say) not only in physics and chemistry but also in biology after his first chemotherapy treatment. But the cancer did not leave him alone. Perhaps due to the high-intensity work of preparing for the celebration of his 40th anniversary at Caltech, he was way too exhausted and his cancer recurred. He then had to receive treatment for a second time. All of us expected that he would take a longer rest this time, but to our surprise, we saw him again in his office within approximately two months. He looked a bit weak but was still as passionate as he used to be, and he started to work as per normal until he was forced to return to the hospital due to pneumonia. When he was in the office, he usually spent hours discussing experimental data and revising manuscripts with different teams. Just one month before his passing, Dr. Zewail was still working on my graphene paper. We went through several iterations before submitting it, and it was probably the last paper he handled (Photo 53.1). While in hospital, he still fussed about the paper and approved the final revision, even though he could no longer write anything on the manuscript.

Although Dr. Zewail had reached the heights of success in his career and was suffering from serious illness, he never ceased in his desire to push the frontier of science forward. He always encouraged us to strive for a breakthrough. In my first ultrafast electron crystallography team meeting with him, he mentioned two standards on which he based his selection of projects: one is that it has to be interesting and the other is that it has to be significant. Dr. Zewail required us to find and develop projects that were more than just "interesting." Thus, although every member in the Zewail group had the freedom to work on the project that he or she was interested in, he or she also

---

Rippling ultrafast dynamics of suspended monolayer graphene

Jianbo Hu[1], Giovanni M. Vanacore[1], Andrea Cepellotti[2], Nicola Marzari[2], and Ahmed H. Zewail[1,*]

[1]Physical Biology Center for Ultrafast Science and Technology, Arthur Amos Noyes Laboratory of Chemical Physics, California Institute of Technology, Pasadena, CA 91125, U.S.A

[2]Theory and Simulation of Materials (THEOS) and National Centre for Computational Design and Discovery of Novel Materials (MARVEL), École Polytechnique Fédérale de Lausanne, 1015 Lausanne, Switzerland

Photo 53.1. The last paper handled by Dr. Zewail.

had to think harder to satisfy Dr. Zewail's standards, which was actually not an easy job. But in this way, we were well trained to be independent investigators. Once Dr. Zewail approved someone's idea, he would give his full support not only in terms of resources but also in terms of his effort and time. Dr. Zewail's great vision significantly improved the quality of our research.

On August 2, 2016, when the sad news of Dr. Zewail's death spread through the laboratory and beyond, I could not accept this fact for a while because I believed that he had acquired his incredible strength from science to fight against the evil disease that afflicted him. I am stuck at the moments when he was healthy (Photo 53.2).

In those four years, I witnessed the burning out of a great star and saw how a real scientist faces life and death, which was an invaluable lesson for me. In science, life may have an end, but Dr. Zewail's contributions to and passion for science will carry on and continue stimulating me for many years to come.

Photo 53.2. My first group retreat at El Portal restaurant, Pasadena.

54 The Art of Smiling through Time

Bin Chen*

My name is Bin Chen, a postdoctoral scholar in Professor Zewail's group at Caltech since 2014. It has been such a beautiful place for both scientific research and daily life until one day, the world collapsed. On 2 August 2016, I was in the office just after lunch when my colleague, Dr. Mohammed Kaplan, came in and told me that Professor Zewail was no longer with us (he showed me the website where he saw the news). My mind went blank after I heard the sad news. I had just met with him the month before to discuss a collaboration with Professor Luis Bañares, and it was hard to believe that he was just gone like that, even till now. Professor Zewail was always so enthusiastic, happy, and devoted to his work. He always had the curiosity, creativity, passion, humor, and energy to drive his ideas forward. I still remember vividly his cheerful and upbeat spirit. His spirit remains forever with us. It is really difficult to express my feelings in a few words. I will take the opportunity here to share several stories of my lovely and respected advisor, Professor Zewail.

Before I came to Caltech, I worked in the field of *in situ* electron microscopy including transmission and scanning electron microscopy. During that period, the time resolution for observing the dynamics of materials by these *in situ* techniques was in the range of

* Postdoctoral Research Fellow.
Email: cbcce@sjtu.edu.cn.

seconds to hours. I dreamed that one day I would have the chance to study ultrafast materials dynamics with the temporal resolution down to femtoseconds using ultrafast electron microscopy. As my contract at the National Institute for Materials Science came to a close, I contacted Professor Zewail in January 2014 to request for a postdoctoral position. The next morning, I got a reply from him when I opened my mailbox. He wrote, "In principle, I am happy to have you in my group…" I cannot express with just a few words how excited I was to see this message, but he definitely made my dream come true. Later on, I found out that he had just partially recovered from an illness, but that did not stop him from devoting time to work as soon as he possibly could because of his true passion and love of science.

I will always remember my first meeting with Professor Zewail. The meeting took place at his office on a hot and sunny day in May 2014 and I was deeply eager to meet the Nobel laureate. After we sat down, he looked at my curriculum vitae printed by our secretary, De Ann Lewis, and asked me several questions. During the whole conversation, although I maintained a smile on my face, I honestly felt a little bit nervous inside. That was my first time talking directly to a Nobel laureate. He also smiled most of the time which helped calm my nerves. Although the conversation was not that long, I got the sense that Caltech was going to be an amazing place where I could do science with such a wise advisor (see group Photo 54.1). I was so touched and impressed because he was gracious with a good sense of humor. Finally, he said to me before I left his office, "I like your smile, please keep your smile." I suddenly realized that smiling was such a simple but key ingredient for establishing friendships, passion, and trust in both daily life and scientific research.

The process of preparing and discussing manuscripts was always enjoyable with Professor Zewail. In May 2016, my colleague, Dr. Xuewen Fu, drafted a manuscript on liquid-cell 4D electron microscopy. This was the first report ever prepared on materials dynamics in liquids at the nanosecond-nanometer spatiotemporal resolution. I clearly remembered that Professor Zewail was very excited when this project was completed. He once proudly introduced this project to the Caltech president during his laboratory tour.

Photo 54.1. This photograph was taken in 2014 at the Athenaeum, Caltech, celebrating Thanksgiving Day. First row, from left to right: Sang Tae Park, Ahmed H. Zewail, Jau Tang, Jongweon Cho, Maggie Sabanpan, Byung-Kuk Yoo; second row, from left to right: De Ann Lewis, Mohammed Hassan, Haihua Liu, Jianbo Hu, Bin Chen, Zixue Su, Xuewen Fu; third row, from left to right: Anthony W. P. Fitzpatrick, Ebrahim Najafi, Giovanni M. Vanacore, Wenxi Liang.

During the manuscript preparation, he said to us, "I would like to publish the exciting results in very nice journals." We knew later that this meant that the manuscript would be sent to pre-eminent journals like *Science* or *Nature*. We were all impressed that he was both meticulous and prompt in offering his comments and suggestions, even with corrections of single words or punctuations. We were also touched and grateful that he was willing to work on our manuscript until very late at night (3.00 a.m. thereabouts!) even though he had not fully recovered from the illness. From this experience, we learnt an important lesson on writing scientific reports. With regards to some unclear expressions in the first draft, he told us: "Take the fog out of your brain." He emphasized that it was very important to make everything clear so that other groups could reproduce the results.

There was a particular episode that occurred during the manuscript discussion which would forever encourage me to push forward in every step and moment. On that day of the manuscript discussion, I parked my car along San Pasqual Street near Caltech. Typically, there is a two-hour limit for free street parking in Pasadena city. In the heat of our manuscript discussion, we were all so engrossed that we forgot about the time until my cellphone suddenly vibrated. This was an alarm reminding me to move my car; otherwise, I would receive a fine ticket. I was completely unaware that our discussion had already lasted two hours. I then explained my situation to Professor Zewail and kindly asked to leave for several minutes. When he heard this, he smiled and made some jokes. Despite the long, intense discussion, we all relaxed because of these jokes. Later on, he always fondly told the same warm and amicable joke at the beginning of group seminars: "Bin, have you moved your car?" I would answer with a smile: "Yes, Sir." This message is short and simple, but it always gives me the reminder, encouragement, hope, and happiness to go forward in my scientific career.

You will forever be alive in my heart, my lovely and respected advisor, Professor Zewail. You were always open and optimistic with a good sense of humor. I appreciate very much all the lessons you have taught me and will remember all the moments we shared, both at and outside of work. The journey is a voyage through time, with the art of smiling.

Bin Chen received his B.S. degree from Wuhan University of Technology in 2003 and his M.S. degree from Tsinghua University in 2006. In 2009, he received his Ph.D. degree in materials science and engineering at the University of Tsukuba. He then held postdoctoral positions in Japan and Australia with a focus on the *in situ* electron microscopy investigation of materials. From 2014 to 2017, he was postdoctoral scholar at the California Institute of Technology where he investigated the ultrafast dynamics of semiconductor materials using 4D electron microscopy. He is now an associate professor in the School of Chemistry and Chemical Engineering at the Shanghai Jiao Tong University. His research interests focus on the development and application of 4D electron microscopy and other *in situ* microscopy techniques with high spatiotemporal resolutions for the study of (ultrafast) dynamics of matter.

55 Never Limit Your Mind — Memorial for Professor Zewail

Xuewen Fu*

I joined Professor Zewail's group as a postdoctoral fellow on November 24, 2014 after I got my Ph.D. from Peking University. It felt both coincidental and predestined for me to have the opportunity to work with Professor Zewail before he passed away. In the third year of my Ph.D., one of my senior laboratory mates passed me a book titled *4D Electron Microscopy — Imaging in Space and Time* when he graduated and said, "You might use it one day in the future!" At the beginning, I didn't pay much attention to this book because my work, which was mainly focused on ultrafast exciton dynamics of nanostructural semiconductors by optical spectroscopy, had little to do with 4D electron microscopy. Therefore, I only had a glance at the cover of this book and found the author was "Ahmed H. Zewail from Caltech." That was the first time I had come across the person known as Professor Zewail, the 1999 Nobel laureate for chemistry due to his pioneering work in femtochemistry. I had never thought I would have the opportunity to work with Professor Zewail because of our completely different research fields. However, in the fourth year of my Ph.D., due to a special requirement of my project, I got an accidental opportunity to work at the École Polytechnique Fédérale de Lausanne in Switzerland with time-resolved cathodeluminescence which was based on ultrafast

* Postdoctoral Research Fellow.
Email: xuewenf@gmail.com; xfu@bnl.gov.

electron pulse techniques. Thus, I began to investigate the literature on the development of ultrafast electron microscopy (UEM) and found that Professor Zewail pioneered 4D electron microscopy after he got the Nobel Prize and his group was the most prestigious one in this field. After finishing my project in Switzerland, I became interested in the ultrafast electron pulse technique and began dreaming of working in Professor Zewail's group on 4D electron microscopy one day.

At the end of my Ph.D. in 2014, I sent an email to Professor Zewail to apply for a postdoctoral position in his group. However, I did not get a reply after nearly two months of waiting and resigned myself to the possibility that I was no longer under consideration. In the middle of the night on May 26, 2014, when I was at my most disheartened, I suddenly received a call from the U.S. and a woman's voice passed through the phone, "Hello, this is De Ann from Dr. Zewail's group. Is this Dr. Fu?" I answered yes and she said, "Please hold on, Dr. Zewail wants to speak with you." I was so surprised and heard Professor Zewail's voice, "Dr. Fu, thank you for your interest in our group. We have finished the evaluation of your application and decided to give you an offer to work as a postdoc at Caltech. Would you like to accept our offer?" I was shocked and immediately answered, "Yes, I'd like to take your offer." "That's great, our secretary will contact you to proceed with your visa application. Now you can have a good sleep," replied Professor Zewail. Just like that, without any interview. I had successfully joined Professor Zewail's group and realized my dream.

As I vividly recall, I joined Professor Zewail's group on November 24, 2014 and had the great honor of participating in the annual Thanksgiving party just after my arrival. This was the first time I met Professor Zewail and got to know all the members of the group. It was also an impressive party in celebration of science. The party segment which impressed me the most was the game "Guess who got the Nobel Prize" hosted by Dr. Jau. He would describe the scientific research of a Nobel laureate and others had to guess who this Nobel Prize winner was. From this game, I could feel that the passion for scientific research runs deep in this group and also found Professor Zewail to be a very jovial scientist. The whole group took a very nice photo at the end of the party (see Photo 55.1), which is my first photo in the Zewail group.

After the Thanksgiving party, I had my first discussion with Professor Zewail about my research plans. He said, "Dr. Fu, welcome to our group and I hope you'll enjoy the sunshine of California! Congratulations on becoming an independent postdoctoral researcher in the fantastic field of 4D electron microscopy. Based on your background, I would like to put you in the UEM1 Lab and you will work with Dr. Bin Chen. I will not restrict your research projects and you can try any of your own ideas with our fantastic UEM instrument. Don't impose any limitations on your ideas and projects; if you get any new thoughts, try to realize them. I will await your exciting experimental data in the office." I was deeply inspired by Professor Zewail's words as he gave me the freedom to do research in this group and encouraged me to proceed with examining my ideas.

I spent nearly half a year studying the features and manipulation of the UEM1 system. In the process, I discovered that the primary

Photo 55.1. Group photograph taken on Thanksgiving Day, 2014.

advantage of UEM1 was its promising ability to do single-shot imaging with nanometer-nanosecond time resolution. Therefore, I proposed several ideas, such as exploring the rotation dynamics of DNA, laser cooling dynamics of semiconductors, and development of liquid cell 4D electron microscopy, and discussed the feasibility of these ideas with Professor Zewail. Professor Zewail's scientific intuition is second to none. After giving these ideas some thought, he suggested that I should focus on the development of liquid cell 4D electron microscopy. Professor Zewail was not only a prestigious scientist but also a respectable advisor and collaborator. As we walked out together from the meeting room after the chat, Professor Zewail put his arm over my shoulder and said encouragingly, "Don't be limited by any previous work. Do your best to try any method you think is possible to get the liquid into our 4D electron microscopy." However, just as I was about to start this project, the UEM1 laser system broke down and all the projects in the UEM1 laboratory had to be stopped. When we reported this bad news to Professor Zewail, he said jokingly, "Dr. Fu and Dr. Chen, you guys broke down my precious UEM1 laser, so now you owe me a *Science* paper." Even though it was meant as a joke by Professor Zewail, I sure felt some pressure as I failed to get any valuable data in the first half a year of my postdoctoral work. After about six months, we acquired a new laser for UEM1 and I resumed the liquid cell 4D electron microscopy project. After two months of hard work, I successfully resolved the technique for the preparation of liquid cells and got it integrated into our 4D electron microscopy. However, just as my project was starting to progress, Professor Zewail's health worsened and he had to get treatment in the hospital for at least two months. During this period, we still kept in touch with him by email and phone. Following his suggestions, I proceeded to finish the first experimental 4D imaging of rotational dynamics of gold nanoparticles in liquid and reported the results to Professor Zewail. He was very excited; upon his arrival back from the hospital, he asked us to come by his office to have a meeting and discuss the data. After the discussion, he said, "The results are great. It is time for you guys to organize the data and prepare a manuscript as soon as possible." As I remembered, in the second meeting, we showed four figures to

Professor Zewail and he gave many suggestions on how the figures might be improved. He then asked us which journal we would like to have this work published in. One of our coauthors, Dr. Hassan, proposed *Nature Communications*, and both Bin and I agreed. Instead, Professor Zewail said, "Why not consider *Science*? I think we have a fifty percent chance of getting it published there."

As per Professor Zewail's suggestions, we spent about 10 days preparing the first draft in the format expected by *Science*. I was quite impressed by the subsequent revisions of our manuscript with Professor Zewail. He was truly a hardworking scientist and was serious with each and every manuscript produced from his group. Every time we sent Professor Zewail our revised manuscript, he was always very efficient and called us to his office for further discussion the very next day. Especially for the last revision after going through more than ten rounds, he asked us to his office at 8.00 a.m. and said, "Dr. Fu, I finished the revision of your manuscript at 3.00 a.m. in the middle of the night. I think we are close to the end. Today, we will do the last revision together as I may have some concerns which require your interpretation." Therefore, we just sat beside Professor Zewail in his office and worked together with him. He carefully read the whole manuscript sentence by sentence, and whenever he had a question, he stopped and asked me to address his question. As I remembered, we worked together until 5.00 p.m. that day. When we finally completed the revision, Professor Zewail said humorously, "Thanks a lot for the cooperation gentlemen, you can proceed with the submission tomorrow. We have worked closely for almost ten days; I don't like to see you guys anymore." All of us laughed and he gave each of us a piece of chocolate when we left his office. The next day, on June 24, 2016, we submitted our manuscript to *Science*. During the next group meeting, Professor Zewail asked me with a smile, "Dr. Fu, how do you like our efficiency?" I smiled and said, "It was amazing and quite impressive!" This was the last time I saw his smile and heard his voice.

After about one month, we received comments from reviewers for our manuscript. Our manuscript was rejected, but the editor offered a resubmission opportunity because of the importance and novelty of our research topic. We were asked to provide more experimental data and conduct a deeper physical model analysis in the revised manuscript. Professor Zewail forwarded the review comments to us and asked us to give a summarized response, after which he would decide how to proceed next. This was the last email I had from him. After about one week, Professor Zewail passed away and left us forever. The whole group was stunned and saddened for the next several months, but we didn't give up on the resubmission opportunity given by the editor; it was the only and best way we could honor the shared effort between Professor Zewail and us. Therefore, we spent over a month getting more data and proposing a complete physical model for our new discoveries. With the major revision, our manuscript was significantly improved and finally accepted by *Science*. Even though we did our best and finally realized Professor Zewail's aspirations, we did not have the chance to share our good news with him. I will miss Professor Zewail forever and will always remember and be inspired by his passion for scientific research in my future academic career.

56 Macroscopic Coherence

Mohammed Kaplan*

In his Wolf Prize acceptance address, Zewail wrote, "How wonderful and significant would it be if humans learn such *peaceful* coexistence between molecules and use the human energy toward the benefit of mankind." These words reflect two main criteria in Zewail's personality, these being (1) his enthusiasm for science and (2) humanity's welfare. Having his roots in his beloved Egypt, he was very much concerned with the advancement of science in Egypt, the Middle East, and the developing world in general.

My first contact with Dr. Zewail was on the phone one October night in 2015 to discuss the possibility of me joining his group as a postdoctoral researcher after finishing my Ph.D. He was very enthusiastic and optimistic about applying 4-dimensional ultrafast electron microscopy to biological systems. "Enthusiasm" and "optimism" were two inherent characteristics of Dr. Zewail; with the sky being the limit, he used to say, "I would love to have a breakthrough in the lab everyday."

The first personal meeting I had with Dr. Zewail was in January 2016 when I visited his group to give a talk about my Ph.D. work. I was initially very nervous at the prospect of giving a presentation in front of him. However, after I met him, his modesty and sincerity

---

* Postdoctoral Research Fellow.
Email: mohammed@caltech.edu

relieved much of my stress. Knowing that I am originally from Iraq, he told me about his visit to Iraq in the 1970s where he was offered a faculty position (and a villa on the Euphrates!) in Baghdad. The visit was part of a trip he did in the Middle East (including Iraq, Lebanon, and Egypt) after his postdoctoral work as he searched for a suitable faculty position within that region. He always wanted to utilise the state-of-the-art knowledge that he had acquired to improve the situation in the Middle East. However, due to the "bureaucracy" there which hampered any scientific achievement, he decided to join Caltech instead which was a decision he would never regret!

In those brief discussions that I was fortunate enough to have with him during my short stay in his group, he used to express his sadness about the current situation in the Middle East, which has an abundance of "human and natural resources that are wasted by absurd wars." It pained him that a region with such great cultural heritage would have such a poor impact on scientific developments in today's "age of science." He hoped that one day all the people in that region can work in a "coherent" way, just like molecules, to exploit its great human potential which will then benefit the whole region and ultimately the whole world.

Finally, I would like to mention that a great dream of Dr. Zewail was to establish a scientific base in the Middle East that will one day become an attractive point for all the smart minds there. To achieve this, he embarked on building the Zewail City in Egypt and worked very hard to make it happen and hoped to see it one day becoming the "Caltech of the Middle East." Although I was not involved in that project, I witnessed the great effort he put in to make Zewail City grow to meet the standards of the developed world. I hope that this City, as a legacy of Dr. Zewail, will thrive and become a source of light and knowledge in Egypt, the Middle East, and the whole world.

Mohammed Kaplan holds an M.D. degree from Iraq. He did his M.Sc. in molecular and cellular life sciences at Utrecht University. After that, he pursued his Ph.D. studies in solid-state nuclear magnetic resonance and dynamic nuclear polarisation, also at Utrecht University. In 2016, he joined Caltech as a postdoctoral researcher to work on the biological applications of ultrafast electron microscopy. Currently, he is working on cryo-electron microscopy at the same institute.

Curriculum Vitae

Ahmed H. Zewail

California Institute of Technology (Caltech)
Arthur Amos Noyes Laboratory of Chemical Physics, Mail Code 127-72
1200 East California Boulevard, Pasadena, California 91125, U.S.A.

	Linus Pauling Chair Professor of Chemistry & Professor of Physics		**Director Physical Biology Center for Ultrafast Science & Technology**
Telephone:	(626) 395-6536	(direct)	(626) 395-2345
FAX:	(626) 792-8456	(direct)	(626) 796-8315
Assistants:	(626) 395-6516		(626) 395-2611

Email: zewail@caltech.edu
Home Page: http://www.zewail.caltech.edu

PERSONAL

Nationality: Egyptian and American
Place of birth: Damanhur, Egypt
Family status: Married, four children

ACADEMIC POSITIONS

Linus Pauling Professor of Chemistry and Professor of Physics, Caltech (1995–)
Director, Physical Biology Center for Ultrafast Science & Technology, Caltech (2005–)
Director, NSF Laboratory for Molecular Sciences, Caltech (1996–2007)
Linus Pauling Professor of Chemical Physics, Caltech (1990–1994)
Professor of Chemical Physics, Caltech (1982–1989)
Associate Professor of Chemical Physics, Caltech (1978–1982)
Assistant Professor of Chemical Physics, Caltech (1976–1978)

IBM Postdoctoral Fellow, University of California, Berkeley (1974–1976)
Predoctoral Research Fellow, University of Pennsylvania (1970–1974)
Teaching Assistant, University of Pennsylvania (1969–1970)
Instructor and Researcher, Alexandria University (1967–1969)
Undergraduate Trainee, Shell Corporation, Alexandria (1966)

ACADEMIC DEGREES

Alexandria University, Egypt (1967); B.S., First Class Honors
Alexandria University, Egypt (1969); M.S.
University of Pennsylvania, Philadelphia, Pennsylvania, U.S.A. (1974); Ph.D.

HONORARY DEGREES

Oxford University, U.K. (1991); M.A., h.c. (Arts)
American University in Cairo, Egypt (1993); D.Sc., h.c. (Science)
Katholieke Universiteit, Leuven, Belgium (1997); D.Sc., h.c. (Science)
University of Pennsylvania, Philadelphia, Pennsylvania, U.S.A. (1997); D.Sc., h.c. (Science)
Université de Lausanne, Switzerland (1997); D.Sc., h.c. (Science)
Swinburne University of Technology, Melbourne, Australia (1999); D.U., h.c. (University)
Arab Academy for Science, Technology and Maritime Transport, Alexandria, Egypt (1999); H.D.A.Sc. (Appl. Science)
Alexandria University, Egypt (1999); D.Ph., h.c. (Philosophy)
University of New Brunswick, Fredericton, Canada (2000); D.Sc., h.c. (Science)
University of Rome "La Sapienza", Italy (2000); D., h.c. (Doctor)
Université de Liège, Belgium (2000); D., h.c. (Doctor)
Queen of Angels-Hollywood Presbyterian Medical Center, Los Angeles, California, U.S.A. (2000); Honorary Member of the Medical Staff, M.D. (Medicine)
Jadavpur University, Kolkata, India (2001); D.Sc., h.c. (Science)
Concordia University, Montréal, Canada (2002); L.L.D., h.c. (Law)

Heriot-Watt University, Edinburgh, Scotland (2002); D.Sc., h.c. (Science)
Pusan National University, Busan, Korea (2003); M.D., h.c. (Medicine)
Lund University, Sweden (2003); D.Ph., h.c. (Philosophy)
Bo gaziçi University, Istanbul, Turkey (2003); D.Sc., h.c. (Science)
École Normale Supérieure, Paris, France (2003); D.Sc., h.c. (Science)
Oxford University, U.K. (2004); D.Sc., h.c. (Science)
Peking University, P. R. China (2004); H.D.D. (University)
Autonomous University of the State of Mexico, Toluca, Mexico (2004); D., h.c. (University)
Trinity College, University of Dublin, Ireland (2004); D.Sc., h.c. (Science)
Tohoku University, Sendai, Japan (2005); H.D.D. (University)
American University of Beirut, Lebanon (2005); D.H.L. (Humane Letters)
University of Buenos Aires, Argentina (2005); D., h.c. (University)
National University of Cordoba, Argentina (2005); D., h.c. (University)
Cambridge University, U.K. (2006); D.Sc., h.c. (Science)
Babes-Bolyai University, Cluj, Romania (2006); H.D.D. (University)
Mansoura University, Mansoura, Egypt (2007); D.Sc., h.c. (Science)
Universiti Malaya, Kuala Lumpur, Malaysia (2007); D.Sc., h.c. (Science)
Universiti Teknologi Kuala Lumpur, Malaysia (2007); D.Sc., h.c. (Science)
Universidad Complutense de Madrid, Spain (2008); D., h.c. (Doctor)
University of Jordan, Amman, Kingdom of Jordan (2009); D.Sc., h.c. (Arts and Sciences)
École Polytechnique Fédérale de Lausanne, Switzerland (2009); D.Sc., h.c. (Science)
Southwestern University, Austin, Texas, U.S.A. (2010); D.H.L. (Humane Letters)
National Chiao Tung University, Hsinchu, Taiwan (2010); D., h.c. (University)
Lead City University, Nigeria (2010); D., h.c. (University)
Eötvös Loránd University, ELTE, Budapest, Hungary (2010); D., h.c. (Natural Sciences)
Boston University, Boston, Massachusetts, U.S.A. (2011); D.Sc., h.c. (Science)
University of Glasgow, U.K. (2011); D.Sc., h.c. (Science)

Baku State University, Baku, Azerbaijan (2011); H.D.D. (University)
University of Tunis, El Manar, Tunisia (2012); D., h.c. (University)
Yale University, New Haven, Connecticut, U.S.A. (2014); D.Sc., h.c. (Science)
University of York, U.K. (2014); H.D.D. (University)
Simon Fraser University, Vancouver, Canada (2014); D.Sc., h.c. (Science)

PROFESSORSHIPS

University of Amsterdam, John van Geuns Stichting Visiting Professor, Netherlands (1979)
University of Bordeaux, Visiting Professor, Paris, France (1981)
École Normale Supérieure, Visiting Professor, France (1983)
University of Kuwait, Visiting Professor, Kuwait (1987)
University of California, Visiting Scholar, Los Angeles, California, U.S.A. (1988)
American University in Cairo, Distinguished Visiting Professor, Egypt (1988)
Johann Wolfgang Goethe-Universität, Rolf Sammet Professor, Frankfurt, Germany (1990)
Oxford University, Sir Cyril Hinshelwood Chair, Visiting Professor, Oxford, U.K. (1991)
Texas A&M University, Visiting Professor, College Station, Texas, U.S.A. (1992)
University of Iowa, Visiting Professor, Iowa City, Iowa, U.S.A. (1992)
Collège de France, Visiting Professor, Paris, France (1995)
Katholieke Universiteit, Visiting Professor, Leuven, Belgium (1998)
University of Würzburg, Röntgen Visiting Professor, Germany (1999)
Université de Lausanne, Honorary Chair Professor, Switzerland (2000)
University of Cambridge, Linnett Professor, U.K. (2002)
United Nations University, Distinguished Chair of Science & Technology Policy, Tokyo, Japan (2003–)
Huazhong University, Honorary Professor, Wuhan, P.R. China (2004–)
Fudan University, Honorary Professor, Shanghai, P.R. China (2004–)
École Normale Supérieure, Blaise Pascal Honorary Professor, Paris, France (2004–2005)

Tohoku University, First Honorary University Professor, Sendai, Japan (2005)
University System of Taiwan, Honorary Chair, Taipei, Taiwan (2010–2020)
Alexandria University, Distinguished Professor, Egypt (2010–)
Eötvös Loránd University, ELTE, Honorary Professor, Budapest, Hungary (2010–)
Imperial College London, Visiting Honorary Professor, UK (2015–)

PRESIDENTIAL AND OFFICIAL APPOINTMENTS

President Obama's Council of Advisors on Science and Technology (2009–2013)
President of Egypt's Advisory Council of Distinguished Scholars and Experts (2014–)
UN Secretary General Ban Ki-moon's Scientific Advisory Board (2013–)
United States First Science Envoy to the Middle East (2009–2011)

SPECIAL HONORS

King Faisal International Prize in Science, King Faisal Foundation (1989)
First Linus Pauling Chair, Caltech (1990)
Femtochemistry Conferences; Solvay, Nobel, and International Series (1993–)
Wolf Prize in Chemistry, Wolf Foundation (1993)
Order of Merit, conferred by the President of Egypt (1995)
Leonardo Da Vinci Award of Excellence, Moët Hennessy—Louis Vuitton Foundation (1995)
Robert A. Welch Award in Chemistry, Welch Foundation (1997)
Benjamin Franklin Medal, Franklin Institute, Philadelphia, Pennsylvania, U.S.A. (1998)
Nobel Prize in Chemistry, Nobel Foundation (1999)
Order of the Grand Collar of the Nile, Highest State Honor, conferred by the President of Egypt (1999)
Order of Zayed, Highest Presidential Honor, U.A.E. (2000)

Order of Cedar, Highest Rank of Commander, conferred by the President of Lebanon (2000)
Order of ISESCO, First Class, conferred by the Prince of Saudi Arabia Salman Ibn Abdel Aziz (2000)
Order of Merit, Highest State Honor, conferred by the President of Tunisia (2000)
Insignia of Pontifical Academy, conferred by the Pope John Paul II (2000)
Order of the Two Niles, First Class, Highest State Honor, conferred by the President of Sudan (2004)
Albert Einstein World Award, World Cultural Council (2006)
Cowl Hood, honor of the Coptic Orthodox Church, Bucharest, Romania (2010)
National Leadership Award, Merage Foundation (2010)
Priestley Gold Medal, highest award, American Chemical Society (2011)
Top American Leaders Award, Washington Post and Harvard University (2011)
Ordre national de la Légion d'honneur, Chevalier, decreed by President of France (2012)

PUBLIC RECOGNITION

Postage Stamps issued in Egypt: "The Portrait" (1998); "The Fourth Pyramid" (1999)
Dr. Ahmed Zewail High School, Disuq, Egypt (1998)
Dr. Ahmed Zewail Medan (Square), Alexandria, Egypt (2000)
Ahmed Zewail streets and schools in Damanhur, Disuq, Cairo and other cities, Egypt (2000–)
Ahmed Zewail Fellowships, University of Pennsylvania, Philadelphia, Pennsylvania, U.S.A. (2000–)
Ahmed Zewail Prize, American University in Cairo, Egypt (2001–)
Exhibition, Nobel Museum, Stockholm, Sweden (2001)
BBC Documentary (2001)
Postage Stamp, issued in Ghana (2002)
Ahmed Zewail Center for FemtoScience Technology, Korea (2002)

Dr. Ahmed Zewail Prize for Creativity in the Arts, Opera Culture Center, Cairo, Egypt (2004–)
Zewail Foundation for Knowledge and Development, Cairo, Egypt (2004–)
Ahmed Zewail Prizes for Excellence and Leadership, International Centre for Theoretical Physics, Trieste, Italy (2004)
Film Documentary by Turk Pipkin, "Nobelity" (2005)
Ahmed Zewail Award for Ultrafast Science and Technology, American Chemical Society (2005–)
Ahmed Zewail Prize in Molecular Sciences, Elsevier (2006–)
Documentary, "A.-Z. Alphabet of Time", Slovak Television (2007)
Ahmed H. Zewail Gold Medal Award & Distinguished Lecture, Wayne State University, Detroit, Michigan, U.S.A. (2008–)
Ahmed Zewail Lecture Hall, Alexandria University, Egypt (2010)
Zewail Foundation Lecture Series, American University in Cairo, Egypt (2011–)
Zewail City of Science and Technology, Cairo, Egypt, Founder (2011–)
Thought Leaders 2014, Top 100 Global, Tages Anzeiger, Switzerland (2014)
Thought Leaders 2015, Top 100 Global, Tages Anzeiger, Switzerland (2015)

SELECTED AWARDS AND PRIZES

Fellow, Alfred P. Sloan Foundation (1978–1982)
Camille and Henry Dreyfus Teacher-Scholar Award, Camille and Henry Dreyfus Foundation (1979–1985)
Alexander von Humboldt Award for Senior United States Scientists, Alexander von Humboldt Foundation (1983)
Award for Especially Creative Research, National Science Foundation (1984; 1988; 1993; 2014)
Buck-Whitney Medal, American Chemical Society (1985)
Fellow, John Simon Guggenheim Memorial Foundation (1987)
Harrison Howe Award, American Chemical Society (1989)
International Award, Carl Zeiss, Germany (1992)

Earle K. Plyler Prize, American Physical Society (1993)
Medal of the Royal Netherlands Academy of Arts and Sciences, Holland (1993)
Bonner Chemiepreis, University of Bonn, Germany (1994)
Herbert P. Broida Prize, American Physical Society (1995)
Collége de France Medal, Paris, France (1995)
Peter Debye Award, American Chemical Society (1996)
Chemical Sciences Award, National Academy of Sciences, U.S.A. (1996)
J. G. Kirkwood Medal, Yale University, New Heaven, Connecticut, U.S.A. (1996)
Peking University Medal, conferred by the PU President, Beijing, P. R. China (1996)
Pittsburgh Spectroscopy Award, Spectroscopy Society of Pittsburgh, Philadelphia, U.S.A. (1997)
First E. B. Wilson Award, American Chemical Society (1997)
Linus Pauling Medal Award, American Chemical Society (1997)
Richard C. Tolman Medal Award, American Chemical Society (1998)
William H. Nichols Medal Award, American Chemical Society (1998)
Paul Karrer Gold Medal, University of Zürich, Switzerland (1998)
E. O. Lawrence Award, U.S. Government (1998)
Merski Award, University of Nebraska, Lincoln, NB (1999)
Röntgen Prize, (100th Anniversary of The Discovery of X-rays), Germany (1999)
Faye Robiner Award, Ross University School of Medicine, New York City, New York, U.S.A. (2000)
Golden Plate Award, American Academy of Achievement (2000)
City of Pisa Medal, conferred by the Mayor of Pisa, Italy (2000)
Medal of "La Sapienza" ("Wisdom"), University of Rome, Italy (2000)
Médaille de l'Institut du Monde Arabe, Paris, France (2000)
Honorary Medal, Université du Centre, Monastir, Tunisia (2000)
Honorary Medal, conferred by the Mayor of Monastir, Tunisia (2000)
Distinguished Alumni Award, University of Pennsylvania, Philadelphia, Pennsylvania, U.S.A. (2002)
G. M. Kosolapoff Award, American Chemical Society (2002)
Distinguished American Service Award, American Arab Anti-Discrimination Committee (2002)

Sir C. V. Raman Award, Indian Institute of Science Education and Research, Kolkata, India (2002)
Arab American Award, National Museum, Dearborn, Michigan, U.S.A. (2004)
Gold Medal (Highest Honor), Burgos University, Burgos, Spain (2004)
Medal, Slovak Academy of Science (2005)
Gold Medal, Slovak Chemical Society (2005)
Grand Gold Medal, Comenius University, Bratislava, Slovak Republic (2005)
Medal of University of Buenos Aires, Argentina (2005)
Medal of National University of Cordoba, Argentina (2005)
Jubilee Medal, National Research Council of Egypt (2006)
Linus Pauling Medal Award, Stanford University, Stanford, California, U.S.A. (2007)
Award, American Medical Students Association (2007)
Inaugural Award of MSE, Stanford University, Stanford, California, U.S.A. (2007)
150th Anniversary Medal of the French Chemical Society, Paris (2007)
Gold Jubilee Medal (50th Anniversary), Assiut University, Assiut, Egypt (2007)
Jabir Ibn Hayyan ("Geber") Medal, Chemical Society, Saudi Arabia (2008)
MIT Lifetime Achievement Award, Arab Students' Organization, Massachusetts Institute of Technology, Cambridge, Massachusetts, U.S.A. (2008)
700th Anniversary Medal, Universidad Complutense de Madrid, Spain (2008)
Othmer Gold Medal, Chemical Heritage Foundation (2009)
Arab American of the Year Award, Arab Community Center for Economic and Social Services, Dearborn, Michigan, U.S.A. (2010)
Award, Sayling Wen Foundation (2010)
Medal of the City of Istanbul, "555 Year Memento of the Ottoman Empire", conferred by the Mayor of Istanbul, Turkey (2010)
Pioneer in Photonics Award, Duke University, Durham, North Carolina, U.S.A. (2010)

375th Anniversary Celebration Medal, Eötvös Loránd University, ELTE, Budapest, Hungary (2010)
G. Robert Oppenheimer Medal, Los Alamos (2010)
Gilbert Newton Lewis Medal, University of California at Berkeley (2010)
President's Medal, American University in Cairo, Egypt (2011)
Al Ahram Key, Al Ahram Foundation, Cairo, Egypt (2011)
Sir Humphrey Davy Medal, Royal Society of London, U.K. (2011)
Sven Berggren Prize, Royal Physiographic (Natural Science) Society, Lund, Sweden (2011)
Honorary Award Medal, Baku State University, Baku, Azerbaijan (2011)
Medal of the University of Tunis, El Manar, Tunisia (2012)
Mendel Medal, Villanova University, Villanova, Pennsylvania (2012)
World Harmony Award, University of California at Santa Barbara (2012)
Shield of Suez Canal Authority, Egypt (2014)
Shield of Zewail City University Student Union, Egypt (2014)
Transformative Achievement Award, Society for Design and Process Science, Fort Worth, Texas (2015)

ACADEMIES AND SOCIETIES

Fellow, American Physical Society (elected 1982)
Member, National Academy of Sciences, U.S.A. (elected 1989)
Member, Third World Academy of Sciences, Italy (elected 1989)
Fellow, St. Catherine's College, Oxford, U.K. (elected 1991)
Member, Sigma Xi Society (elected 1992); Honorary Fellow (2009–)
Member, American Academy of Arts and Sciences (elected 1993)
Member, Académie Européenne des Sciences, des Arts et des Lettres, France (elected 1994)
Member, American Philosophical Society (elected 1998)
Member, Pontifical Academy of Sciences, Vatican (elected 1999)
Member, American Academy of Achievement (elected 1999)
Member, Royal Danish Academy of Sciences & Letters (elected 2000)
Fellow, American Association for the Advancement of Science (elected 2000)

Honorary Fellow, Chemical Society of India (elected 2001)
Member, Indian Academy of Sciences (elected 2001)
Foreign Member, Royal Society of London, U.K. (elected 2001)
Honorary Fellow, Sydney Sussex College, Cambridge, U.K. (elected 2002)
Foreign Fellow, Indian National Science Academy (elected 2002)
Honorary Foreign Member, Korean Academy of Science and Technology (elected 2002)
Honorary Fellow, African Academy of Sciences, Kenya (elected 2002)
Honorary Fellow, Royal Society of Chemistry, U.K. (elected 2003)
Foreign Member, Russian Academy of Sciences (elected 2003)
Foreign Member, Royal Swedish Academy of Sciences (elected 2003)
Foreign Member, Royal Academy of Belgium (elected 2003)
Honorary Fellow, Islamic World Academy of Sciences, Jordan (elected 2003)
Honorary Fellow, St. Catherine's College, Oxford, U.K. (elected 2004)
Honorary Member, European Academy of Sciences, Belgium (elected 2004)
Honorary Fellow, Literary & Historical Society, University College, Dublin, Ireland (elected 2004)
Honorary Member, Board of Advisors, National Society of High School Scholars, U.S.A. (elected 2004)
Founding Member, Academia Bibliotheca Alexandrinae, Egypt (2004)
Foreign Member, French Academy of Sciences (elected 2005)
Honorary Fellow with PM Mahathir Mohamed and PM Ahmad Badawi, Academy of Sciences Malaysia (elected 2005; inducted 2007)
Foreign Member, Royal Spanish Academy of Sciences (elected 2007; inducted 2008)
Member, Chinese Academy of Sciences (elected 2009)
Honorary Member, American Chamber of Commerce (elected 2010)
Foreign Fellow, Nigerian Academy of Science (elected 2010)
Fellow, American Chemical Society (elected 2010)
Honorary President, Science Age Society, *Asr Al Alm*, Egypt (2010–)
Honorary President, Asian Nanoscience and Nanotechnology Association, Taiwan (2010–)

Member, Royal Physiographic (Natural Science) Society, Sweden (elected 2012)

Member, Académie tunisienne des sciences, des lettres et des arts (Beït Al-Hikma), Tunisia (elected 2012)

Foreign Member, Azerbaijan National Academy of Sciences (elected 2012)

BOARDS AND ADVISORY SERVICES

Member of Advisory and Editorial Boards; Editor of scientific journals and book series; Chairman and Member of Organizing Committees of national and international conferences; Member of Boards including the following:

Chief Editor, Chemical Physics Letters (1991–2007); Honorary Advisory Editor (2007–)

Member, Board of Advisors, World Scientific, Singapore (1994–)

Member, Board of Advisors, Max Planck Institute, Germany (1994–)

Member, Board of Trustees, American University in Cairo, Egypt (1999–)

Member, Board of Trustees, Bibliotheca Alexandria (2001–2005); Board of Advisors (2006)

Member, Scientific Advisory Board, Welch Foundation (2002–)

Patron, Multilateral Initiative on Malaria (2003–)

Member, Board of Directors, Qatar Foundation (2003–2011)

Member, Scientific Advisory Board, Chalmers University of Technology, Göteborg, Sweden (2003–)

Member, Board of Trustees, TIAA-CREF, U.S.A. (2004–2007)

Honorary Member, Board of Trustees, Future University, Sudan (2004–)

Member, Board of Advisors, Nanyang Technological University, Singapore (2005–)

Member, Scientific Council, École Normale Supérieure, Paris, France (2005–)

Co-Chair, High Level (Blue Ribbon) Amiri Panel, Kuwait (2007)

Member, International Advisory Board, King Abdullah University of Science & Technology, Jeddah, Saudi Arabia (2007)

Honorary Member, National Advisory Board, Arab American National Museum (2007–)

President, International Jury, L'Oréal-UNESCO "For Women in Science" Awards (2007–)

Member, Supreme Council for Science and Technology, Egypt (2007–)

President, 44th International Chemistry Olympiad, American Chemical Society (2010–2012)

Lifetime Member, President's Council, New York Academy of Sciences (2010–)

Member, UNESCO High Panel on Science, Technology, and Innovation for Development, Paris, France (2011–)

Chairman, Board of Trustees, Zewail City of Science and Technology, Cairo, Egypt (2011–)

Member, International Solvay Institutes, Brussels, Belgium (2011–)

Member, 21st Century Council, Global Thinkers, Nicolas Berggruen Institute, Beverly Hills, California, U.S.A. (2011–)

Member, Federation of American Scientists, Board of Sponsors, Washington, D.C., U.S.A. (2012–)

PUBLICATIONS AND PRESENTATIONS

Articles

Some 600 articles (authored and co-authored) have been published in the fields of science, education, and world affairs.

Books

(1) *Advances in Laser Spectroscopy I*, ed. A. H. Zewail, SPIE, Bellingham, 1977

(2) *Advances in Laser Chemistry*, ed. A. H. Zewail, Springer-Verlag, Berlin-Heidelberg, 1978

(3) *Photochemistry and Photobiology, Vols. 1 and 2*, ed. A. H. Zewail, Harwood Academic, London, 1983

(4) *Ultrafast Phenomena VII*, eds. C. B. Harris, E. P. Ippen, G. A. Mourou and A. H. Zewail, Springer-Verlag, Berlin-Heidelberg, 1990

(5) *The Chemical Bond: Structure and Dynamics*, ed. A. H. Zewail, Academic Press, Boston, 1992

(6) *Ultrafast Phenomena VIII*, eds. J.-L. Martin, A. Migus, G. A. Mourou and A. H. Zewail, Springer-Verlag, Berlin-Heidelberg, 1993

(7) *Ultrafast Phenomena IX*, eds. P. F. Barbara, W. H. Knox, G. A. Mourou and A. H. Zewail, Springer-Verlag, Berlin-Heidelberg, 1994

(8) *Femtochemistry: Ultrafast Dynamics of the Chemical Bond*, A. H. Zewail, *Vols. 1 and 2*, World Scientific, Singapore, 1994

(9) *Voyage Through Time: Walks of Life to the Nobel Prize*, A. H. Zewail, American University in Cairo (AUC), Cairo, 2002; appeared in 19 languages and editions: English, French, German, Spanish, Romanian, Hungarian, Russian, Arabic, Chinese, Korean, Bahasa Malaysia, Indonesian, Hindi, and Azerbaijani

(10) *Age of Science* (*Asr Al Álm*, in Arabic), A. H. Zewail, Dar Al Shorouk, Beirut-Cairo, 2005; appeared in the 15th edition since publication in June 2005

(11) *Time* (*Al Zaman*, in Arabic), Book Series, A. H. Zewail, Dar Al Shorouk, Cairo, 2007

(12) *Dialogue of Civilizations* (*Hewar Al Hadarat*, in Arabic), Book Series, A. H. Zewail, Dar Al Shorouk, Cairo, 2007

(13) *Physical Biology: From Atoms to Medicine*, ed. A. H. Zewail, Imperial College Press, London, 2008

(14) *4D Electron Microscopy: Imaging in Space and Time*, A. H. Zewail and J. M. Thomas, Imperial College Press, London, 2010

(15) *4D Visualization of Matter: Recent Collected Works*, A. H. Zewail, Imperial College Press, London, 2014

(16) *Reflections on World Affairs: Peace and Politics*, A. H. Zewail, Imperial College Press, London, 2015

www.ingramcontent.com/pod-product-compliance
Lightning Source LLC
Chambersburg PA
CBHW061232220326
41599CB00028B/5397
* 9 7 8 9 8 1 3 2 3 1 6 5 8 *